Encyclopaedia of

Tools and Techniques in Biochemistry and Molecular Biology

Encyclopaedia of

Tools and Techniques in Biochemistry and Molecular Biology

Volume 1

Editor:

Rashmi Rana

AGROTECH PRESS

JAIPUR ▪ NEW DELHI

AGROTECH PRESS

Regd. Off. "Bony Residency", Gate No. 2
Opp. Tilak Public School, Vishweswariya Nagar,
Gopalpura Road, Jaipur - 302018, India
Ph: 0141 - 2761280, Telefax: 2761381
E-mail : agrotechpress@gmail.com

Delhi Office: 102, 1st Floor, Satyam House,
4327/3, Ansari Road, Darya Ganj,
New Delhi - 110002, India
Ph: 011-45652440
visit us: www.agrotechpress.com

Distributed By
Keya International Press LLP
E-mail: keyainternationalpressllp@gmail.com

Published in India, 2025

ISBN : 9789384568245 (4 Vols. Set)

Typeset by Thomson Press

Printed and bound in India by Thomson Press Ltd.,
New Delhi

Contents

Preface

Molecular biology is the study of biology at a molecular level. The field overlaps with other areas of biology and chemistry, particularly genetics and biochemistry. Molecular biology chiefly concerns itself with understanding the interactions between the various systems of a cell, including the interactions between DNA, RNA and protein biosynthesis as well as learning how these interactions are regulated.

Researchers in molecular biology use specific techniques native to molecular biology, but increasingly combine these with techniques and ideas from genetics and biochemistry. There is not a defined line between these disciplines. "Biochemistry" is the study of the chemical substances and vital processes occurring in living organisms. Biochemists focus heavily on the role, function, and structure of biomolecules. The study of the chemistry behind biological processes and the synthesis of biologically active molecules are examples of biochemistry. "Genetics" is the study of the effect of genetic differences on organisms. Often this can be inferred by the absence of a normal component. The study of "mutants" – organisms which lack one or more functional components with respect to the so-called "wild type" or normal phenotype. Genetic interactions can often confound simple interpretations of such "knock-out" studies. "Molecular biology" is the study of molecular underpinnings of the process of replication, transcription and translation of the genetic material. The central dogma of molecular biology where genetic material is transcribed into RNA and then translated into protein, despite being an oversimplified picture of molecular biology, still provides a good starting point for understanding the field. This picture, however, is undergoing revision in light of emerging novel roles for RNA. Much of the work in molecular biology is quantitative, and recently much work has been done at the interface of molecular biology and computer science in bioinformatics and computational biology. As of the early 2000s, the study of gene structure and function, molecular genetics, has been amongst the most prominent sub-field of molecular biology. Increasingly many other loops of biology focus on molecules, either directly studying their interactions in their own right such as in cell biology and developmental biology, or indirectly, where the techniques of molecular biology are used to infer historical attributes of populations or species, as in fields in evolutionary biology such as population genetics and phylogenetics. There is also a long tradition of studying biomolecules "from the ground up" in biophysics. An overview of the history of molecular biology revealed the original convergence of geneticists, physicists, and structural chemists on a common problem: the structure and function of the gene. Conceptual and methodological frameworks from each of these disciplinary strands united in the ultimate determination of the double helical structure of DNA along with the mechanisms of gene replication, mutation, and expression. With this recent history in mind, philosophers of molecular biology have examined the key concepts of the field: *mechanism*, *information*, and *gene*. Moreover, molecular biology has provided cases for addressing more general issues in the philosophy of science, such as reduction and the integration of fields, and the nature of explanation without laws of nature in biology. It has been argued that, given the importance of the discovery of macromolecular mechanisms throughout the history of molecular biology, a philosophical focus on mechanisms generates the clearest picture of its history, of its concepts, and of the cases from its past utilized by philosophers of science. Over the last 40 years biochemistry has become so successful at explaining living processes that now almost all areas of the life sciences from botany to medicine are engaged in biochemical research. Today the main focus of pure biochemistry is in understanding how biological molecules give rise to the processes that occur within living cells, which in turn relates greatly to the study and understanding of whole organisms.

This is a very impressive well-organised, beautifully-illustrated introductory textbook. It should prove popular among a very broad variety of students for whom understanding of principles is more important than memorization of large amounts of poorly comprehended factual information. There are many such students around the world who will benefit from this masterly book.

—Editor

1

Biochemistry: Catalysts of Life

Enzymes: The Catalysts of Life

Each cell in the body is a beehive of activity called metabolism. The biochemical activity is the process of chemical and physical change which goes on continually in the living organism. These characteristics of life include the build-up of new tissue, replacement of old tissue, conversion of food to energy, disposal of waste materials, and reproduction.

This building up and tearing down takes place in the face of an apparent paradox. The greatest majority of these biochemical reactions do not take place spontaneously. The phenomenon of catalysis makes possible biochemical reactions necessary for all life processes. Catalysis is defined as the acceleration of a chemical reaction by some substance which itself undergoes no permanent chemical change.

The catalysts of biochemical reactions are enzymes and are responsible for bringing about almost all of the chemical reactions in living organisms. Without enzymes, these reactions take place at a rate far too slow for the pace of metabolism.

The oxidation of a fatty acid to carbon dioxide and water is not a gentle process in a test tube – extremes of pH, high temperatures and corrosive chemicals are required. Yet in the body, such a reaction takes place smoothly and rapidly within a narrow range of pH and temperature. In the laboratory, the average protein must be boiled for about 24 hours in a 20% HCl solution to achieve a complete breakdown. In the body, the breakdown takes place in four hours or less under conditions of mild physiological temperature and pH.

It is through attempts at understanding more about enzyme catalysts – what they are, what they do, and how they do it– that many advances in medicine and the life sciences have been brought about. All known enzymes are proteins. They are high molecular weight compounds made up principally of chains of amino acids linked together by peptide bonds. The primary benefit of enzymes for us is in digestion...

Digestive Enzymes

Specific enzymes work on specific foods. You need the right type of enzyme for the foods you want it to break down. Digestive enzymes are enzymes that break down food into usable material. The major different types of digestive enzymes are:

- amylase – breaks down carbohydrates, starches, and sugars which are prevalent in potatoes, fruits, vegetables, and many snack foods
- protease – breaks down proteins found in meats, nuts, eggs, and cheese
- lipase – breaks down fats found in most dairy products, nuts, oils, and meat
- cellulase – breaks down cellulose, plant fiber; not found in humans

Enzyme is a Biological Catalyst

Catalyst is a molecule that increases the rate of a chemical reaction without itself being permanently changed after the reaction. Enzymes are proteins that catalyse (*i.e.*, increase or decrease the rates of) chemical reactions. In enzymatic reactions, the molecules at the beginning of the process are called substrates, and they are converted into different molecules, called the products. Almost all processes in a biological cell need enzymes to occur at significant rates. Since enzymes are selective for their substrates and speed up only a few reactions from among many possibilities, the set of enzymes made in a cell determines which metabolic pathways occur in that cell.

Like all catalysts, enzymes work by lowering the activation energy ($E_a^{\ddagger}$) for a reaction, thus dramatically increasing the rate of the reaction. As a result, products are formed faster and reactions reach their equilibrium state more rapidly. Most enzyme reaction rates are millions of times faster than those of comparable un-catalysed reactions. As with all catalysts, enzymes are not consumed by the reactions they catalyse, nor do they alter the equilibrium of these reactions.

However, enzymes do differ from most other catalysts by being much more specific. Enzymes are known to catalyse about 4,000 biochemical reactions. A few RNA molecules called ribozymes also catalyse reactions, with an important example being some parts of the ribosome. Synthetic molecules called artificial enzymes also display enzyme-like catalysis. Enzyme activity can be affected by other molecules. Inhibitors are molecules that decrease enzyme activity; activators are molecules that increase activity.

Many drugs and poisons are enzyme inhibitors. Activity is also affected by temperature, chemical environment (*e.g.*, pH), and the concentration of substrate. Some enzymes are used commercially, for example, in the synthesis of antibiotics. In addition, some household products use enzymes to speed up biochemical reactions (*e.g.*, enzymes in biological washing powders break down protein or fat stains on clothes; enzymes in meat tenderizers break down proteins, making the meat easier to chew).

Structures and Mechanisms

Enzymes are generally globular proteins and range from just 62 amino acid residues in size, for the monomer of 4-oxalocrotonate tautomerase, to over 2,500 residues in the animal fatty acid synthase. A small number of RNA-based biological catalysts exist, with the most common being the ribosome; these are referred to as either RNA-enzymes or ribozymes.

The activities of enzymes are determined by their three-dimensional structure. However, although structure does determine function, predicting a novel enzyme's activity just from its structure is a very difficult problem that has not yet been solved.

Most enzymes are much larger than the substrates they act on, and only a small portion of the enzyme (around 3–4 amino acids) is directly involved in catalysis. The region that contains these catalytic residues, binds the substrate, and then carries out the reaction is known as the active site. Enzymes can also contain sites that bind cofactors, which are needed for catalysis. Some enzymes also have binding sites for small molecules, which are often direct or indirect products or substrates of the reaction catalysed. This binding can serve to increase or decrease the enzyme's activity, providing a means for feedback regulation.

Like all proteins, enzymes are long, linear chains of amino acids that fold to produce a three-dimensional product. Each unique amino acid sequence produces a specific structure, which has unique properties. Individual protein chains may sometimes group together to form a protein complex.

Most enzymes can be denatured—that is, unfolded and inactivated—by heating or chemical denaturants, which disrupt the three-dimensional structure of the protein. Depending on the enzyme, denaturation may be reversible or irreversible. Structures of enzymes in complex with substrates or substrate analogs during a reaction may be obtained using Time resolved crystallography methods.

Specificity

Enzymes are usually very specific as to which reactions they catalyse and the substrates that are involved in these reactions. Complementary shape, charge and hydrophilic/hydrophobic characteristics of enzymes and substrates are responsible for this specificity. Enzymes can also show impressive levels of stereospecificity, regioselectivity and chemoselectivity.

Some of the enzymes showing the highest specificity and accuracy are involved in the copying and expression of the genome. These enzymes have "proof-reading" mechanisms. Here, an enzyme such as DNA polymerase catalyzes a reaction in a first step and then checks that the product is correct in a second step.

This two-step process results in average error rates of less than 1 error in 100 million reactions in high-fidelity mammalian polymerases. Similar proofreading mechanisms are also found in RNA polymerase, aminoacyl tRNA synthetases and ribosomes.

Some enzymes that produce secondary metabolites are described as promiscuous, as they can act on a relatively broad range of different substrates. It has been suggested that this broad substrate specificity is important for the evolution of new biosynthetic pathways.

"Lock and Key" Model

Enzymes are very specific, and it was suggested by the Nobel laureate organic chemist Emil Fischer in 1894 that this was because both the enzyme and the substrate possess specific complementary geometric shapes that fit exactly into one another.

This is often referred to as "the lock and key" model. However, while this model explains enzyme specificity, it fails to explain the stabilization of the transition state that enzymes achieve.

In 1958, Daniel Koshland suggested a modification to the lock and key model: since enzymes are rather flexible structures, the active site is continually reshaped by interactions with the substrate as the substrate interacts with the enzyme. As a result, the substrate does not simply bind to a rigid active site; the amino acid side chains which make up the active site are molded into the precise positions that enable the enzyme to perform its catalytic function.

In some cases, such as glycosidases, the substrate molecule also changes shape slightly as it enters the active site. The active site continues to change until the substrate is completely bound, at which point the final shape and charge is determined. Induced fit may enhance the fidelity of molecular recognition in the presence of competition and noise via the conformational proofreading mechanism.

Mechanisms

Enzymes can act in several ways, all of which lower $\Delta G^{\ddagger}$:

- Lowering the activation energy by creating an environment in which the transition state is stabilized (e.g. straining the shape of a substrate—by binding the transition-state conformation of the substrate/product molecules, the enzyme distorts the bound substrate(s) into their transition state form, thereby reducing the amount of energy required to complete the transition).
- Lowering the energy of the transition state, but without distorting the substrate, by creating an environment with the opposite charge distribution to that of the transition state.
- Providing an alternative pathway. For example, temporarily reacting with the substrate to form an intermediate ES complex, which would be impossible in the absence of the enzyme.
- Reducing the reaction entropy change by bringing substrates together in the correct orientation to react. Considering $\Delta H^{\ddagger}$ alone overlooks this effect.
- Increases in temperatures speed up reactions. Thus, temperature increases help the enzyme function and develop the end product even faster. However, if heated too much, the enzyme's shape deteriorates and only when the temperature comes back to normal does the enzyme regain its shape. Some enzymes like thermolabile enzymes work best at low temperatures.

Interestingly, this entropic effect involves destabilization of the ground state, and its contribution to catalysis is relatively small.

Transition State Stabilization

The understanding of the origin of the reduction of $\Delta G^{\ddagger}$ requires one to find out how the enzymes can stabilize its transition state more than the transition state of the uncatalyzed reaction. Apparently, the most effective way for reaching large stabilization is the use of electrostatic effects, in particular, by having a relatively fixed polar environment that is oriented toward the charge distribution of the transition state. Such an environment does not exist in the uncatalyzed reaction in water.

Dynamics and Function

The internal dynamics of enzymes is linked to their mechanism of catalysis. Internal dynamics are the movement of parts of the enzyme's structure, such as individual amino acid residues, a group of amino acids, or even an entire protein domain. These movements occur at various time-scales ranging from femtoseconds to seconds. Networks of protein residues throughout an enzyme's structure can contribute to catalysis through dynamic motions. Protein motions are vital to many enzymes, but whether small and fast vibrations, or larger and slower conformational movements are more important depends on the type of reaction involved.

However, although these movements are important in binding and releasing substrates and products, it is not clear if protein movements help to accelerate the chemical steps in enzymatic reactions. These new insights also have implications in understanding allosteric effects and developing new drugs.

Allosteric Modulation

Allosteric sites are sites on the enzyme that bind to molecules in the cellular environment. The sites form weak, noncovalent bonds with these molecules, causing a change in the conformation of the enzyme. This change in conformation translates to the active site, which then affects the reaction rate of the enzyme. Allosteric interactions can both inhibit and activate enzymes and are a common way that enzymes are controlled in the body.

Cofactors and Coenzymes

Cofactors

Some enzymes do not need any additional components to show full activity. However, others require non-protein molecules called cofactors to be bound for activity. Cofactors can be either inorganic (*e.g.*, metal ions and iron-sulphur clusters) or organic compounds (e.g., flavin and heme). Organic cofactors can be either prosthetic groups, which are tightly bound to an enzyme, or coenzymes, which are released from the enzyme's active site during the reaction. Coenzymes include NADH, NADPH and adenosine triphosphate. These molecules transfer chemical groups between enzymes. An example of an enzyme that contains a cofactor is carbonic anhydrase, and is shown in the ribbon diagram above with a zinc cofactor bound as part of its active site. These tightly bound molecules are usually found in the active site and are involved in catalysis. For example, flavin and heme cofactors are often involved in redox reactions.

Enzymes that require a cofactor but do not have one bound are called *apoenzymes* or *apoproteins*. An apoenzyme together with its cofactor(s) is called a *holoenzyme* (this is the active form). Most cofactors are not covalently attached to an enzyme, but are very tightly bound. However, organic prosthetic groups can be covalently bound (*e.g.*, thiamine pyrophosphate in the enzyme pyruvate dehydrogenase). The term "holoenzyme" can also be applied to enzymes that contain multiple protein subunits, such as the DNA polymerases; here the holoenzyme is the complete complex containing all the subunits needed for activity.

Coenzymes

Coenzymes are small organic molecules that transport chemical groups from one enzyme to another. Some of these chemicals such as riboflavin, thiamine and folic acid are vitamins (compounds which cannot be synthesized by the body and must be acquired from the diet). The chemical groups carried include the hydride ion (H^-) carried by NAD or $NADP^+$, the phosphate group carried by adenosine triphosphate, the acetyl group carried by coenzyme A, formyl, methenyl or methyl groups carried by folic acid and the methyl group carried by S-adenosylmethionine. Since coenzymes are chemically changed as a consequence of enzyme action, it is useful to consider coenzymes to be a special class of substrates, or second substrates, which are common to many different enzymes.

For example, about 700 enzymes are known to use the coenzyme NADH. Coenzymes are usually continuously regenerated and their concentrations maintained at a steady level inside the cell: for example, NADPH is regenerated through the pentose phosphate pathway and *S*-adenosylmethionine by methionine adenosyltransferase. This continuous regeneration means that even small amounts of coenzymes are used very intensively. For example, the human body turns over its own weight in ATP each day.

Kinetics

Enzyme kinetics is the investigation of how enzymes bind substrates and turn them into products. The rate data used in kinetic analyses are obtained from enzyme assays.

In 1902 Victor Henri proposed a quantitative theory of enzyme kinetics, but his experimental data were not useful because the significance of the hydrogen ion concentration was not yet appreciated. After Peter Lauritz Sørensen had defined the logarithmic pH-scale and introduced the concept of buffering in 1909 the German chemist Leonor Michaelis and his Canadian postdoc Maud Leonora Menten repeated Henri's experiments and confirmed his equation which is referred to as Henri-Michaelis-Menten kinetics (sometimes also Michaelis-Menten kinetics).

Their work was further developed by G. E. Briggs and J. B. S. Haldane, who derived kinetic equations that are still widely used today. The major contribution of Henri was to think of enzyme reactions in two stages. In the first, the substrate binds reversibly to the enzyme, forming the enzyme-substrate complex.

This is sometimes called the Michaelis complex. The enzyme then catalyzes the chemical step in the reaction and releases the product. Enzymes can catalyse up to several million reactions per second. For example, the uncatalyzed decarboxylation of orotidine 5'-monophosphate has a half life of 78 million years. However, when the enzyme orotidine 5'-phosphate decarboxylase is added, the same process takes just 25 milliseconds. Enzyme rates depend on solution conditions and substrate concentration. Conditions that denature the protein abolish enzyme activity, such as high temperatures, extremes of pH or high salt concentrations, while raising substrate concentration tends to increase activity. To find the maximum speed of an enzymatic reaction, the substrate concentration is increased until a constant rate of product formation is seen.

This is shown in the saturation curve on the right. Saturation happens because, as substrate concentration increases, more and more of the free enzyme is converted into the substrate-bound ES form. At the maximum velocity (V_{max}) of the enzyme, all the enzyme active sites are bound to substrate, and the amount of ES complex is the same as the total amount of enzyme. However, V_{max} is only one kinetic constant of enzymes. The amount of substrate needed to achieve a given rate of reaction is also important.

This is given by the Michaelis-Menten constant (K_m), which is the substrate concentration required for an enzyme to reach one-half its maximum velocity. Each enzyme has a characteristic K_m for a given substrate, and this can show how tight the binding of the substrate is to the enzyme. Another useful constant is k_{cat}, which is the number of substrate molecules handled by one active site per second.

The efficiency of an enzyme can be expressed in terms of k_{cat}/K_m. This is also called the specificity constant and incorporates the rate constants for all steps in the reaction. Because the specificity constant reflects both affinity and catalytic ability, it is useful for comparing different enzymes against each other, or the same enzyme with different substrates.

The theoretical maximum for the specificity constant is called the diffusion limit and is about 10^8 to 10^9 (M^{-1} s^{-1}). At this point every collision of the enzyme with its substrate will result in catalysis, and the rate of product formation is not limited by the reaction rate but by the diffusion rate. Enzymes with this property are called *catalytically perfect* or *kinetically perfect*. Example of such enzymes are triose-phosphate isomerase, carbonic anhydrase, acetylcholinesterase, catalase, fumarase, β-lactamase, and superoxide dismutase.

Michaelis-Menten kinetics relies on the law of mass action, which is derived from the assumptions of free diffusion and thermodynamically driven random collision. However, many biochemical or cellular processes deviate significantly from these conditions, because of macromolecular crowding, phase-separation of the enzyme/substrate/product, or one or two-dimensional molecular movement. In these situations, a fractal Michaelis-Menten kinetics may be applied.

Some enzymes operate with kinetics which are faster than diffusion rates, which would seem to be impossible. Several mechanisms have been invoked to explain this phenomenon. Some proteins are believed to accelerate catalysis by drawing their substrate in and pre-orienting them by using dipolar electric fields. Other models invoke a quantum-mechanical tunnelling explanation, whereby a proton or an electron can tunnel through activation barriers, although for proton tunnelling this model remains somewhat controversial. Quantum tunnelling for protons has been observed in tryptamine. This suggests that enzyme catalysis may be more accurately characterized as "through the barrier" rather than the traditional model, which requires substrates to go "over" a lowered energy barrier.

Inhibition

Enzyme inhibitors are molecules that bind to enzymes and decrease their activity. Since blocking an enzyme's activity can kill a pathogen or correct a metabolic imbalance, many drugs are enzyme inhibitors. They are also used as herbicides and pesticides. Not all molecules that bind to enzymes are inhibitors; *enzyme activators* bind to enzymes and increase their enzymatic activity.

The binding of an inhibitor can stop a substrate from entering the enzyme's active site and/or hinder the enzyme from catalysing its reaction. Inhibitor binding is either reversible or irreversible. Irreversible inhibitors usually react with the enzyme and change it chemically. These inhibitors modify key amino acid residues needed for enzymatic activity. In contrast, reversible inhibitors bind non-covalently and different types of inhibition are produced depending on whether these inhibitors bind the enzyme, the enzyme-substrate complex, or both.

Many drug molecules are enzyme inhibitors, so their discovery and improvement is an active area of research in biochemistry and pharmacology. A medicinal enzyme inhibitor is often judged by its specificity (its lack of binding to other proteins) and its potency (its dissociation constant, which indicates the concentration needed to inhibit the enzyme). A high specificity and potency ensure that a drug will have few side effects and thus low toxicity.

Enzyme inhibitors also occur naturally and are involved in the regulation of metabolism. For example, enzymes in a metabolic pathway can be inhibited by downstream products. This type of negative feedback slows flux through a pathway when the products begin to build up and is an important way to maintain homeostasis in a cell. Other cellular enzyme inhibitors are proteins that specifically bind to and inhibit an enzyme target. This can help control enzymes that may be damaging to a cell, such as proteases or nucleases; a well-characterised example is the ribonuclease inhibitor, which binds to ribonucleases in one of the tightest known protein–protein interactions. Natural enzyme inhibitors can also be poisons and are used as defences against predators or as ways of killing prey.

Reversible Inhibitors

Types of Reversible Inhibitors

Reversible inhibitors bind to enzymes with non-covalent interactions such as hydrogen bonds, hydrophobic interactions and ionic bonds. Multiple weak bonds between the inhibitor and the active site combine to produce strong and specific binding. In contrast to substrates and irreversible inhibitors, reversible inhibitors generally do not undergo chemical reactions when bound to the enzyme and can be easily removed by dilution or dialysis. There are four kinds of reversible enzyme inhibitors. They are classified according to the effect of varying the concentration of the enzyme's substrate on the inhibitor.

- In competitive inhibition, the substrate and inhibitor cannot bind to the enzyme at the same time. This usually results from the inhibitor having an affinity for the active site of an enzyme where the substrate also binds; the substrate and inhibitor *compete* for access to the enzyme's active site. This type of inhibition can be overcome by sufficiently high concentrations of substrate, i.e., by out-competing the inhibitor. Competitive inhibitors are often similar in structure to the real substrate.
- In uncompetitive inhibition, the inhibitor binds only to the substrate-enzyme complex, it should not be confused with non-competitive inhibitors. Both Vmax and Km decrease (maximum velocity decreases as a result of removing activated complex while binding efficiency increases as a result of Le Chatelier's principle).
- In mixed inhibition, the inhibitor can bind to the enzyme at the same time as the enzyme's substrate. However, the binding of the inhibitor affects the binding of the substrate, and vice versa. This type of inhibition can be reduced, but not overcome by increasing concentrations of substrate. Although it is possible for mixed-type inhibitors to bind in the active site, this type of inhibition generally results from an allosteric effect where the inhibitor binds to a different site on an enzyme. Inhibitor binding to this allosteric site changes the conformation (i.e., tertiary structure or three-dimensional shape) of the enzyme so that the affinity of the substrate for the active site is reduced.
- Non-competitive inhibition is a form of mixed inhibition where the binding of the inhibitor to the enzyme reduces its activity but does not affect the binding of substrate. As a result, the extent of inhibition depends only on the concentration of the inhibitor.

Quantitative Description of Reversible Inhibition

Reversible inhibition can be described quantitatively in terms of the inhibitor's binding to the enzyme and to the enzyme–substrate complex, and its effects on the kinetic constants of the enzyme. In the classic Michaelis–Menten scheme below, an enzyme (E) binds to its substrate (S) to form the enzyme–substrate complex ES.

Upon catalysis, this complex breaks down to release product P and free enzyme. The inhibitor (I) can bind to either E or ES with the dissociation constants K_i or K_i', respectively.

When an enzyme has multiple substrates, inhibitors can show different types of inhibition depending on which substrate is considered. This results from the active site containing two different binding sites within the active site, one for each substrate. For example, an inhibitor might compete with substrate A for the first binding site, but be a non-competitive inhibitor with respect to substrate B in the second binding site.

Measuring the Dissociation Constants of a Reversible Inhibitor

As noted above, an enzyme inhibitor is characterised by its two dissociation constants, K_i and K_i', to the enzyme and to the enzyme-substrate complex, respectively.

The enzyme-inhibitor constant K_i can be measured directly by various methods; one extremely accurate method is isothermal titration calorimetry, in which the inhibitor is titrated into a solution of enzyme and the heat released or absorbed is measured.

However, the other dissociation constant K_i' is difficult to measure directly, since the enzyme-substrate complex is short-lived and undergoing a chemical reaction to form the product. Hence, K_i' is usually measured indirectly, by observing the enzyme activity under various substrate and inhibitor concentrations, and fitting the data to a modified Michaelis–Menten equation.

$$V = \frac{V_{max}[S]}{aK_m + \alpha'[S]} = \frac{(1/\alpha')V_{max}[S]}{(\alpha/\alpha')K_m + [S]}$$

where the modifying factors α and α' are defined by the inhibitor concentration and its two dissociation constants

$$\alpha = 1 + \frac{[I]}{K_i}$$

$$\alpha' = 1 + \frac{[I]}{K_i'}.$$

Thus, in the presence of the inhibitor, the enzyme's effective K_m and V_{max} become $(\alpha/\alpha')K_m$ and $(1/\alpha')V_{max}$, respectively. However, the modified Michaelis-Menten equation assumes that binding of the inhibitor to the enzyme has reached equilibrium, which may be a very slow process for inhibitors with sub-nanomolar dissociation constants. In these cases, it is usually more practical to treat the tight-binding inhibitor as an irreversible inhibitor; however, it can still be possible to estimate K_i' kinetically if K_i is measured independently.

The effects of different types of reversible enzyme inhibitors on enzymatic activity can be visualized using graphical representations of the Michaelis–Menten equation, such as Lineweaver–Burk and Eadie-Hofstee plots. For example, in the Lineweaver–Burk plots at the right, the competitive inhibition lines intersect on the *y*-axis, illustrating that such inhibitors do not affect V_{max}. Similarly, the non-competitive inhibition lines intersect on the *x*-axis, showing these inhibitors do not affect K_m. However, it can be difficult to estimate K_i and K_i' accurately from such plots, so it is advisable to estimate these constants using more reliable nonlinear regression methods, as described above.

Special Cases

- The mechanism of partially co mpetitive inhibition is similar to that of non-competitive, except that the EIS complex has catalytic activity, which may be lower or even higher (partially competitive activation) than that of the enzyme–substrate (ES) complex. This inhibition typically displays a lower V_{max}, but an unaffected K_m value.
- Uncompetitive inhibition occurs when the inhibitor binds only to the enzyme–substrate complex, not to the free enzyme; the EIS complex is catalytically inactive. This mode of inhibition is rare and causes a decrease in both V_{max} and the K_m value.
- Substrate and product inhibition is where either the substrate or product of an enzyme reaction inhibit the enzyme's activity. This inhibition may follow the competitive, uncompetitive or mixed patterns. In substrate inhibition there is a progressive decrease in activity at high substrate concentrations. This may indicate the existence of two substrate-binding sites in the enzyme. At low substrate, the high-affinity site is occupied and normal kinetics are followed. However, at higher concentrations, the second inhibitory site becomes occupied, inhibiting the enzyme. Product inhibition is often a regulatory feature in metabolism and can be a form of negative feedback.
- Slow-tight inhibition occurs when the initial enzyme–inhibitor complex EI undergoes isomerisation to a second more tightly held complex, EI*, but the overall inhibition process is reversible. This manifests itself as

slowly increasing enzyme inhibition. Under these conditions, traditional Michaelis–Menten kinetics give a false value for K_i, which is time-dependent. The true value of K_i can be obtained through more complex analysis of the on (k_{on}) and off (k_{off}) rate constants for inhibitor association.

Examples of Reversible Inhibitors

As enzymes have evolved to bind their substrates tightly, and most reversible inhibitors bind in the active site of enzymes, it is unsurprising that some of these inhibitors are strikingly similar in structure to the substrates of their targets. An example of these substrate mimics are the protease inhibitors, a very successful class of antiretroviral drugs used to treat HIV. The structure of ritonavir, a protease inhibitor based on a peptide and containing three peptide bonds, is shown on the right. As this drug resembles the protein that is the substrate of the HIV protease, it competes with this substrate in the enzyme's active site.

Enzyme inhibitors are often designed to mimic the transition state or intermediate of an enzyme-catalysed reaction. This ensures that the inhibitor exploits the transition state stabilising effect of the enzyme, resulting in a better binding affinity (lower K_i) than substrate-based designs. An example of such a transition state inhibitor is the antiviral drug oseltamivir; this drug mimics the planar nature of the ring oxonium ion in the reaction of the viral enzyme neuraminidase.

However, not all inhibitors are based on the structures of substrates. For example, the structure of another HIV protease inhibitor tipranavir is shown on the left. This molecule is not based on a peptide and has no obvious structural similarity to a protein substrate. These non-peptide inhibitors can be more stable than inhibitors containing peptide bonds, because they will not be substrates for peptidases and are less likely to be degraded.

In drug design it is important to consider the concentrations of substrates to which the target enzymes are exposed. For example, some protein kinase inhibitors have chemical structures that are similar to adenosine triphosphate, one of the substrates of these enzymes.

However, drugs that are simple competitive inhibitors will have to compete with the high concentrations of ATP in the cell. Protein kinases can also be inhibited by competition at the binding sites where the kinases interact with their substrate proteins, and most proteins are present inside cells at concentrations much lower than the concentration of ATP. As a consequence, if two protein kinase inhibitors both bind in the active site with similar affinity, but only one has to compete with ATP, then the competitive inhibitor at the protein-binding site will inhibit the enzyme more effectively.

Irreversible Inhibitors

Types of Irreversible Inhibition

Irreversible inhibitors usually covalently modify an enzyme, and inhibition cannot therefore be reversed. Irreversible inhibitors often contain reactive functional groups such as nitrogen mustards, aldehydes, haloalkanes, alkenes, Michael acceptors, phenyl sulfonates, or fluorophosphonates. These electrophilic groups react with amino acid side chains to form covalent adducts. The residues modified are those with side chains containing nucleophiles such as hydroxyl or sulfhydryl groups; these include the amino acids serine (as in DFP, right), cysteine, threonine or tyrosine.

Irreversible inhibition is different from irreversible enzyme inactivation. Irreversible inhibitors are generally specific for one class of enzyme and do not inactivate all proteins; they do not function by destroying protein structure but by specifically altering the active site of their target. For example, extremes of pH or temperature usually cause denaturation of all protein structure, but this is a non-specific effect. Similarly, some non-specific chemical treatments destroy protein structure: for example, heating in concentrated hydrochloric acid will hydrolyse the peptide bonds holding proteins together, releasing free amino acids.

Irreversible inhibitors display time-dependent inhibition and their potency therefore cannot be characterised by an IC_{50} value. This is because the amount of active enzyme at a given concentration of irreversible inhibitor will be different depending on how long the inhibitor is pre-incubated with the enzyme. Instead, $k_{obs}/[I]$ values are used, wherek_{obs} is the observed pseudo-first order rate of inactivation (obtained by plotting the log of % activity vs. time) and $[I]$ is the concentration of inhibitor. The $k_{obs}/[I]$ parameter is valid as long as the inhibitor does not saturate binding with the enzyme (in which case $k_{obs} = k_{inact}$).

Analysis of Irreversible Inhibition

The irreversible inhibitors form a reversible non-covalent complex with the enzyme (EI or ESI) and this then reacts to produce the covalently modified "dead-end complex" EI*. The rate at which EI* is formed is called the inactivation rate or k_{inact}. Since formation of EI may compete with ES, binding of irreversible inhibitors can be prevented by competition either with substrate or with a second, reversible inhibitor. This protection effect is good evidence of a specific reaction of the irreversible inhibitor with the active site.

The binding and inactivation steps of this reaction are investigated by incubating the enzyme with inhibitor and assaying the amount of activity remaining over time. The activity will be decrease in a time-dependent manner, usually following exponential decay. Fitting these data to a rate equation gives the rate of inactivation at this concentration of inhibitor. This is done at several different concentrations of inhibitor. If a reversible EI complex is involved the inactivation rate will be saturable and fitting this curve will give k_{inact} and K_i. Another method that is widely used in these analyses is mass spectrometry. Here, accurate measurement of the mass of the unmodified native enzyme and the inactivated enzyme gives the increase in mass caused by reaction with the inhibitor and shows the stoichiometry of the reaction.

This is usually done using a MALDI-TOF mass spectrometer. In a complementary technique, peptide mass fingerprinting involves digestion of the native and modified protein with a protease such as trypsin. This will produce a set of peptides that can be analysed using a mass spectrometer. The peptide that changes in mass after reaction with the inhibitor will be the one that contains the site of modification.

Special Cases

Not all irreversible inhibitors form covalent adducts with their enzyme targets. Some reversible inhibitors bind so tightly to their target enzyme that they are essentially irreversible. These tight-binding inhibitors may show kinetics similar to covalent irreversible inhibitors. In these cases, some of these inhibitors rapidly bind to the enzyme in a low-affinity EI complex and this then undergoes a slower rearrangement to a very tightly bound EI* complex. This kinetic behaviour is called slow-binding. This slow rearrangement after binding often involves a conformational change as the enzyme "clamps down" around the inhibitor molecule. Examples of slow-binding inhibitors include some important drugs, such methotrexate, allopurinol, and the activated form of acyclovir.

Examples of Irreversible Inhibitors

The enzyme hydrolyses the phosphorus–fluorine bond, but the phosphate residue remains bound to the serine in the active site, deactivating it. Similarly, DFP also reacts with the active site of acetylcholine esterase in the synapses of neurons, and consequently is a potent neurotoxin, with a lethal dose of less than 100 mg.

Suicide inhibition is an unusual type of irreversible inhibition where the enzyme converts the inhibitor into a reactive form in its active site. An example is the inhibitor of polyamine biosynthesis, α-difluoromethylornithine or DFMO, which is an analogue of the amino acid ornithine, and is used to treat African trypanosomiasis (sleeping sickness). Ornithine decarboxylase can catalyse the decarboxylation of DFMO instead of ornithine, as shown above. However, this decarboxylation reaction is followed by the elimination of a fluorine atom, which converts this catalytic intermediate into a conjugated imine, a highly electrophilic species. This reactive form of DFMO then reacts with either a cysteine or lysine residue in the active site to irreversibly inactivate the enzyme.

Discovery and Design of Inhibitors

New drugs are the products of a long drug development process, the first step of which is often the discovery of a new enzyme inhibitor. In the past the only way to discover these new inhibitors was by trial and error: screening huge libraries of compounds against a target enzyme and hoping that some useful leads would emerge. This brute force approach is still successful and has even been extended by combinatorial chemistry approaches that quickly produce large numbers of novel compounds and high-throughput screening technology to rapidly screen these huge chemical libraries for useful inhibitors.

More recently, an alternative approach has been applied: rational drug design uses the three-dimensional structure of an enzyme's active site to predict which molecules might be inhibitors. These predictions are then tested and one of these tested compounds may be a novel inhibitor. This new inhibitor is then used to try to obtain a structure of the enzyme in an inhibitor/enzyme complex to show how the molecule is binding to the active site, allowing changes to be made to the inhibitor to try to optimise binding. This test and improve cycle is then repeated until a sufficiently potent inhibitor is produced. Computer-based methods of predicting the affinity of an inhibitor for an enzyme are also being developed, such as molecular docking and molecular mechanics.

Uses of Inhibitors

Enzyme inhibitors are found in nature and are also designed and produced as part of pharmacology and biochemistry. Natural poisons are often enzyme inhibitors that have evolved to defend a plant or animal against predators. These natural toxins include some of the most poisonous compounds known. Artificial inhibitors are often used as drugs, but can also be insecticides such as malathion, herbicides such as glyphosate, or disinfectants such as triclosan.

Chemotherapy

The most common uses for enzyme inhibitors are as drugs to treat disease. Many of these inhibitors target a human enzyme and aim to correct a pathological condition. However, not all drugs are enzyme inhibitors. Some, such as anti-epileptic drugs, alter enzyme activity by causing more or less of the enzyme to be produced. These effects are called enzyme induction and inhibition and are alterations in gene expression, which is unrelated to the type of enzyme inhibition. Other drugs interact with cellular targets that are not enzymes, such as ion channels or membrane receptors.

An example of a medicinal enzyme inhibitor is sildenafil (Viagra), a common treatment for male erectile dysfunction. This compound is a potent inhibitor of cGMP specific phosphodiesterase type 5, the enzyme that degrades the signalling molecule cyclic guanosine monophosphate. This signalling molecule triggers smooth muscle relaxation and allows blood flow into the corpus cavernosum, which causes an erection. Since the drug decreases the activity of the enzyme that halts the signal, it makes this signal last for a longer period of time.

Folic acid is a substrate of dihydrofolate reductase, an enzyme involved in making nucleotides that is potently inhibited by methotrexate. Methotrexate blocks the action of dihydrofolate reductase and thereby halts the production of nucleotides. This block of nucleotide biosynthesis is more toxic to rapidly growing cells than non-dividing cells, since a rapidly-growing cell has to carry out DNA replication, therefore methotrexate is often used in cancer chemotherapy.

Drugs also are used to inhibit enzymes needed for the survival of pathogens. For example, bacteria are surrounded by a thick cell wall made of a net-like polymer called peptidoglycan. Many antibiotics such as penicillin and vancomycin inhibit the enzymes that produce and then cross-link the strands of this polymer together. This causes the cell wall to lose strength and the bacteria to burst.

Drug design is facilitated when an enzyme that is essential to the pathogen's survival is absent or very different in humans. In the example above, humans do not make peptidoglycan, therefore inhibitors of this process are selectively toxic to bacteria. Selective toxicity is also produced in antibiotics by exploiting differences in the structure of the ribosomes in bacteria, or how they make fatty acids.

Metabolic Control

Enzyme inhibitors are also important in metabolic control. Many metabolic pathways in the cell are inhibited by metabolites that control enzyme activity through allosteric regulation or substrate inhibition. A good example is the allosteric regulation of the glycolytic pathway. This catabolic pathway consumes glucose and produces ATP, NADH and pyruvate. A key step for the regulation of glycolysis is an early reaction in the pathway catalysed by phosphofructokinase-1 (PFK1). When ATP levels rise, ATP binds an allosteric site in PFK1 to decrease the rate of the enzyme reaction; glycolysis is inhibited and ATP production falls.

This negative feedback control helps maintain a steady concentration of ATP in the cell. However, metabolic pathways are not just regulated through inhibition since enzyme activation is equally important. With respect to PFK1, fructose 2,6-bisphosphate and ADP are examples of metabolites that are allosteric activators. Physiological enzyme inhibition can also be produced by specific protein inhibitors.

This mechanism occurs in the pancreas, which synthesises many digestive precursor enzymes known as zymogens. Many of these are activated by the trypsin protease, so it is important to inhibit the activity of trypsin in the pancreas to prevent the organ from digesting itself. One way in which the activity of trypsin is controlled is the production of a specific and potent trypsin inhibitor protein in the pancreas. This inhibitor binds tightly to trypsin, preventing the trypsin activity that would otherwise be detrimental to the organ. Although the trypsin inhibitor is a protein, it avoids being hydrolysed as a substrate by the protease by excluding water from trypsin's active site and destabilising the transition state. Other examples of physiological enzyme inhibitor proteins include the barstar inhibitor of the bacterial ribonuclease barnase and the inhibitors of protein phosphatases.

Pesticides and Herbicides

Many herbicides and pesticides are enzyme inhibitors. Acetylcholinesterase (AChE) is an enzyme found in animals from insects to humans. It is essential to nerve cell function through its mechanism of breaking down the neurotransmitter acetylcholine into its constituents, acetate and choline. This is somewhat unique among neurotransmitters as most, including serotonin, dopamine, and norepinephrine, are absorbed from the synaptic cleft rather than cleaved.

A large number of AChE inhibitors are used in both medicine and agriculture. Reversible competitive inhibitors, such as edrophonium, physostigmine, and neostigmine, are used in the treatment of myasthenia gravis and in anaesthesia. The carbamate pesticides are also examples of reversible AChE inhibitors. The organophosphate insecticides such as malathion, parathion, and chlorpyrifos irreversibly inhibit acetylcholinesterase.

The herbicide glyphosate is an inhibitor of 3-phosphoshikimate 1-carboxyvinyltransferase, other herbicides, such as the sulfonylureas inhibit the enzyme acetolactate synthase. Both these enzymes are needed for plants to make branched-chain amino acids. Many other enzymess are inhibited by herbicides, including enzymes needed for the biosynthesis of lipids and carotenoids and the processes of photosynthesis and oxidative phosphorylation.

Natural Poisons

Animals and plants have evolved to synthesise a vast array of poisonous products including secondary metabolites, peptides and proteins that can act as inhibitors. Natural toxins are usually small organic molecules and are so diverse that there are probably natural inhibitors for most metabolic processes. The metabolic processes targeted by natural poisons encompass more than enzymes in metabolic pathways and can also include the inhibition of receptor, channel and structural protein functions in a cell. For example, paclitaxel (taxol), an organic molecule found in the Pacific yew tree, binds tightly to tubulin dimers and inhibits their assembly into microtubules in the cytoskeleton.

Many natural poisons act as neurotoxins that can cause paralysis leading to death and have functions for defence against predators or in hunting and capturing prey. Some of these natural inhibitors, despite their toxic attributes, are valuable for therapeutic uses at lower doses. An example of a neurotoxin are the glycoalkaloids, from the plant species in the *Solanaceae* family (includes potato, tomato and eggplant), that are acetylcholinesterase inhibitors. Inhibition of this enzyme causes an uncontrolled increase in the acetylcholine neurotransmitter, muscular paralysis and then death. Neurotoxicity can also result from the inhibition of receptors; for example, atropine from deadly nightshade (*Atropa belladonna*) that functions as a competitive antagonist of the muscarinic acetylcholine receptors.

Although many natural toxins are secondary metabolites, these poisons also include peptides and proteins. An example of a toxic peptide is alpha-amanitin, which is found in relatives of the death cap mushroom. This is a potent enzyme inhibitor, in this case preventing the RNA polymerase II enzyme from transcribing DNA.

The algal toxin microcystin is also a peptide and is an inhibitor of protein phosphatases. This toxin can contaminate water supplies after algal blooms and is a known carcinogen that can also cause acute liver hemorrhage and death at higher doses. Proteins can also be natural poisons or antinutrients, such as the trypsin inhibitors that are found in some legumes. A less common class of toxins are toxic enzymes: These act as irreversible inhibitors of their target enzymes and work by chemically modifying their substrate enzymes.

An example is ricin, an extremely potent protein toxin found in castor oil beans. This enzyme is a glycosidase that inactivates ribosomes. Since ricin is a catalytic irreversible inhibitor, this allows just a single molecule of ricin to kill a cell. Enzyme reaction rates can be decreased by various types of enzyme inhibitors.

Competitive Inhibition

In competitive inhibition, the inhibitor and substrate compete for the enzyme (i.e., they can not bind at the same time). Often competitive inhibitors strongly resemble the real substrate of the enzyme. For example, methotrexate is a competitive inhibitor of the enzyme dihydrofolate reductase, which catalyzes the reduction of dihydrofolate to tetrahydrofolate.

The similarity between the structures of folic acid and this drug. Note that binding of the inhibitor need *not* be to the substrate binding site (as frequently stated), if binding of the inhibitor changes the conformation of the enzyme to prevent substrate binding and *vice versa*. In competitive inhibition the maximal velocity of the reaction is not changed, but higher substrate concentrations are required to reach a given velocity, increasing the apparent K_m.

Uncompetitive Inhibition

In uncompetitive inhibition the inhibitor can not bind to the free enzyme, but only to the ES-complex. The EIS-complex thus formed is enzymatically inactive. This type of inhibition is rare, but may occur in multimeric enzymes.

Non-competitive Inhibition

Non-competitive inhibitors can bind to the enzyme at the same time as the substrate, i.e. they *never* bind to the active site. Both the EI and EIS complexes are enzymatically inactive. Because the inhibitor can not be driven from the enzyme by higher substrate concentration (in contrast to competitive inhibition), the apparent V_{max} changes. But because the substrate can still bind to the enzyme, the K_m stays the same.

Mixed Inhibition

This type of inhibition resembles the non-competitive, except that the EIS-complex has residual enzymatic activity.

In many organisms inhibitors may act as part of a feedback mechanism. If an enzyme produces too much of one substance in the organism, that substance may act as an inhibitor for the enzyme at the beginning of the pathway that produces it, causing production of the substance to slow down or stop when there is sufficient amount. This is a form of negative feedback. Enzymes which are subject to this form of regulation are often multimeric and have allosteric binding sites for regulatory substances. Their substrate/velocity plots are not hyperbolar, but sigmoidal (S-shaped).

Irreversible inhibitors react with the enzyme and form a covalent adduct with the protein. The inactivation is irreversible. These compounds include eflornithine a drug used to treat the parasitic disease sleeping sickness. Penicillin and Aspirin also act in this manner. With these drugs, the compound is bound in the active site and the enzyme then converts the inhibitor into an activated form that reacts irreversibly with one or more amino acid residues.

Uses of Inhibitors

Since inhibitors modulate the function of enzymes they are often used as drugs. An common example of an inhibitor that is used as a drug is aspirin, which inhibits the COX-1 and COX-2 enzymes that produce the inflammation messenger prostaglandin, thus suppressing pain and inflammation. However, other enzyme inhibitors are poisons. For example, the poison cyanide is an irreversible enzyme inhibitor that combines with the copper and iron in the active site of the enzyme cytochrome c oxidase and blocks cellular respiration.

Biological Function

Enzymes serve a wide variety of functions inside living organisms. They are indispensable for signal transduction and cell regulation, often via kinases and phosphatases. They also generate movement, with myosin hydrolysing ATP to generate muscle contraction and also moving cargo around the cell as part of the cytoskeleton. Other ATPases in the cell membrane are ion pumps involved in active transport. Enzymes are also involved in more exotic functions, such as luciferase generating light in fireflies. Viruses can also contain enzymes for infecting cells, such as the HIV integrase and reverse transcriptase, or for viral release from cells, like the influenza virus neuraminidase.

An important function of enzymes is in the digestive systems of animals. Enzymes such as amylases and proteases break down large molecules (starch or proteins, respectively) into smaller ones, so they can be absorbed by the intestines. Starch molecules, for example, are too large to be absorbed from the intestine, but enzymes hydrolyse the starch chains into smaller molecules such as maltose and eventually glucose, which can then be absorbed. Different enzymes digest different food substances. In ruminants which have herbivorous diets, microorganisms in the gut produce another enzyme, cellulase to break down the cellulose cell walls of plant fiber. Several enzymes can work together in a specific order, creating metabolic pathways. In a metabolic pathway, one enzyme takes the product of another enzyme as a substrate. After the catalytic reaction, the product is then passed on to another enzyme. Sometimes more than one enzyme can catalyse the same reaction in parallel, this can allow more complex regulation: with for example a low constant activity being provided by one enzyme but an inducible high activity from a second enzyme.

Enzymes determine what steps occur in these pathways. Without enzymes, metabolism would neither progress through the same steps, nor be fast enough to serve the needs of the cell. Indeed, a metabolic pathway such as glycolysis could not exist independently of enzymes. Glucose, for example, can react directly with ATP to become phosphorylated at one or more of its carbons. In the absence of enzymes, this occurs so slowly as to be insignificant. However, if hexokinase is added, these slow reactions continue to take place except that phosphorylation at carbon 6 occurs so rapidly that if the mixture is tested a short time later, glucose-6-phosphate is found to be the only significant product. Consequently, the network of metabolic pathways within each cell depends on the set of functional enzymes that are present.

Control of Activity

There are five main ways that enzyme activity is controlled in the cell.

1. Enzyme production (transcription and translation of enzyme genes) can be enhanced or diminished by a cell in response to changes in the cell's environment. This form of gene regulation is called enzyme induction and inhibition. For example, bacteria may become resistant to antibiotics such as penicillin because enzymes called beta-lactamases are induced that hydrolyse the crucial beta-lactam ring within the penicillin molecule. Another example are enzymes in the liver called cytochrome P450 oxidases, which are important in drug metabolism. Induction or inhibition of these enzymes can cause drug interactions.
2. Enzymes can be compartmentalized, with different metabolic pathways occurring in different cellular compartments. For example, fatty acids are synthesized by one set of enzymes in the cytosol, endoplasmic reticulum and the Golgi apparatus and used by a different set of enzymes as a source of energy in the mitochondrion, through b-oxidation.
3. Enzymes can be regulated by inhibitors and activators. For example, the end product(s) of a metabolic pathway are often inhibitors for one of the first enzymes of the pathway (usually the first irreversible step, called *committed step*), thus regulating the amount of end product made by the pathways. Such a regulatory mechanism is called a negative feedback mechanism, because the amount of the end product produced is regulated by its own concentration. Negative feedback mechanism can effectively adjust the rate of synthesis of intermediate metabolites according to the demands of the cells. This helps allocate materials and energy economically, and prevents the manufacture of excess end products. The control of enzymatic action helps to maintain a stable internal environment in living organisms.
4. Enzymes can be regulated through post-translational modification. This can include phosphorylation, myristoylation and glycosylation. For example, in the response to insulin, the phosphorylation of multiple enzymes, including glycogen synthase, helps control the synthesis or degradation of glycogen and allows the cell to respond to changes in blood sugar. Another example of post-translational modification is the cleavage of the polypeptide chain. Chymotrypsin, a digestive protease, is produced in inactive form as chymotrypsinogen in the pancreas and transported in this form to the stomach where it is activated. This stops the enzyme from digesting the pancreas or other tissues before it enters the gut. This type of inactive precursor to an enzyme is known as a zymogen.

5. Some enzymes may become activated when localized to a different environment (e.g. from a reducing (cytoplasm) to an oxidizing (periplasm) environment, high pH to low pH etc.). For example, hemagglutinin in the influenza virus is activated by a conformational change caused by the acidic conditions, these occur when it is taken up inside its host cell and enters the lysosome.

Involvement in Disease

Since the tight control of enzyme activity is essential for homeostasis, any malfunction (mutation, overproduction, underproduction or deletion) of a single critical enzyme can lead to a genetic disease. The importance of enzymes is shown by the fact that a lethal illness can be caused by the malfunction of just one type of enzyme out of the thousands of types present in our bodies.

One example is the most common type of phenylketonuria. A mutation of a single amino acid in the enzyme phenylalanine hydroxylase, which catalyzes the first step in the degradation of phenylalanine, results in build-up of phenylalanine and related products. This can lead to mental retardation if the disease is untreated.

Another example is when germline mutations in genes coding for DNA repair enzymes cause hereditary cancer syndromes such as xeroderma pigmentosum. Defects in these enzymes cause cancer since the body is less able to repair mutations in the genome. This causes a slow accumulation of mutations and results in the development of many types of cancer in the sufferer.

Naming Conventions

An enzyme's name is often derived from its substrate or the chemical reaction it catalyzes, with the word ending in *-ase*. Examples are lactase, alcohol dehydrogenase and DNA polymerase. This may result in different enzymes, called isozymes, with the same function having the same basic name. Isoenzymes have a different amino acid sequence and might be distinguished by their optimal pH, kinetic properties or immunologically.

Furthermore, the normal physiological reaction an enzyme catalyzes may not be the same as under artificial conditions. This can result in the same enzyme being identified with two different names. *E.g.* Glucose isomerase, used industrially to convert glucose into the sweetener fructose, is a xylose isomerase *in vivo*.

The International Union of Biochemistry and Molecular Biology have developed a nomenclature for enzymes, the EC numbers; each enzyme is described by a sequence of four numbers preceded by "EC". The first number broadly classifies the enzyme based on its mechanism.

The top-level classification is:

- EC 1 *Oxidoreductases*: catalyse oxidation/reduction reactions
- EC 2 *Transferases*: transfer a functional group (*e.g.* a methyl or phosphate group)
- EC 3 *Hydrolases*: catalyse the hydrolysis of various bonds
- EC 4 *Lyases*: cleave various bonds by means other than hydrolysis and oxidation
- EC 5 *Isomerases*: catalyse isomerization changes within a single molecule
- EC 6 *Ligases*: join two molecules with covalent bonds.

According to the naming conventions, enzymes are generally classified into six main family classes and many sub-family classes. Some web-servers, e.g., EzyPred and bioinformatics tools have been developed to predict which main family class and sub-family class an enzyme molecule belongs to according to its sequence information alone via the pseudo amino acid composition.

Industrial Applications

Enzymes are used in the chemical industry and other industrial applications when extremely specific catalysts are required. However, enzymes in general are limited in the number of reactions they have evolved to catalyse and also by their lack of stability in organic solvents and at high temperatures. Consequently, protein engineering is an active area of research and involves attempts to create new enzymes with novel properties, either through rational design or *in vitro* evolution. These efforts have begun to be successful, and a few enzymes have now been desiged "from scratch" to catalyse reactions that do not occur in nature.

Bioenergetics

Bioenergetics is the part of biochemistry concerned with the energy involved in making and breaking of chemical bonds in the molecules found in biological organisms.

Growth, development and metabolism are some of the central phenomena in the study of biological organisms. The role of energy is fundamental to such biological processes. The ability to harness energy from a variety of metabolic pathways is a property of all living organisms. Life is dependent on energy transformations; living organisms survive because of exchange of energy within and without.

In a living organism, chemical bonds are broken and made as part of the exchange and transformation of energy. Energy is available for work (such as mechanical work) or for other processes (such as chemical synthesis and anabolic processes in growth), when weak bonds are broken and stronger bonds are made. The production of stronger bonds allows release of usable energy. Living organisms obtain energy from organic and inorganic materials. For example, lithotrophs can oxidize minerals such as nitrates or forms of sulphur, such as elemental sulphur, sulfites, and hydrogen sulfide to produce ATP. In photosynthesis, autotrophs can produce ATP using light energy. Heterotrophs must consume organic compounds. These are mostly carbohydrates, fats, and proteins. The amount of energy actually obtained by the organism is lower than the amount present in the food; there are losses in digestion, metabolism, and thermogenesis.

The materials are generally combined with oxygen to release energy, although some can also be oxidized anaerobically by various organisms. The bonds holding the molecules of nutrients together and the bonds holding molecules of free oxygen together are all relatively weak compared with the chemical bonds holding carbon dioxide and water together.

The utilization of these materials is a form of slow combustion. That is why the energy content of food can be estimated with a bomb calorimeter. The materials are oxidized slowly enough that the organisms do not actually produce fire. The oxidation releases energy because stronger bonds have been formed. This net energy may evolve as heat, or some of which may be used by the organism for other purposes, such as breaking other bonds to do chemistry.

Living organisms produce ATP from energy sources via oxidative phosphorylation. The terminal phosphate bonds of ATP are relatively weak compared with the stronger bonds formed when ATP is broken down to adenosine monophosphate and phosphate, dissolved in water. Here it is the energy of hydration that results in energy release. This hydrolysis of ATP is used as a battery to store energy in cells, for intermediate metabolism. Utilization of chemical energy from such molecular bond rearrangement powers biological processes in every biological organism.

Types of Reactions

- Exergonic is a spontaneous reaction that releases energy. It is thermodynamically favoured. On the course of a reaction, energy needs to be put in, this activation energy drives the reactants from a stable state to a highly energetic unstable configuration. These reactants are usually complex molecules that are broken into simpler products. The entire reaction is usually catabolic. The release of energy, also called free energy is a $-\Delta G$ because energy is lost from the bonds formed by the products.
- Endergonic is an anabolic reaction that consumes energy. It has a $+\Delta G$ because energy is required to break bonds.

The free energy (ΔG) gained or lost in a reaction can be calculated: $\Delta G = \Delta H - T\Delta S$.

Also, $\Delta G = \Delta G^{\circ\prime} + 2.303RT\log([P]/[R])$ where:

- o R is the gas constant, 1.987 cal/mol
- o T is temperature in kelvins K = 273 + °C
- o $[P]$ is products
- o $[R]$ is the reactants.

Cotransport

In August 1960, Robert K. Crane presented for the first time his discovery of the sodium-glucose cotransport as the mechanism for intestinal glucose absorption. Crane's discovery of cotransport was the first ever proposal of flux coupling in biology and was the most important event concerning carbohydrate absorption in the 20th century.

Chemiosmotic Theory

One of the major triumphs of bioenergetics is Peter D. Mitchell's chemiosmotic theory of how protons in aqueous solution function in the production of ATP in cell organelles such as mitochondria. Other cellular sources of ATP such as glycolysis were understood first, but such processes for direct coupling of enzyme activity to ATP production are not the major source of useful chemical energy in most cells. Chemiosmotic coupling is the major energy producing process in most cells, being utilized in chloroplasts and several single celled organisms in addition to mitochondria.

Energy Balance

Energy balance is the biological homeostasis of energy in living systems. It is measured with the following equation:

Energy intake = internal heat produced + external work + storage.

Homeostasis

With regards to any given life system parameter, an organism may be a *conformer* or a *regulator*. On one hand, regulators try to maintain the parameter at a constant level over possibly wide ambient environmental variations. On the other hand, conformers allow the environment to determine the parameter. For instance, endothermic animals (mammals and birds) maintain a constant body temperature, while exothermic animals (almost all other organisms) exhibit wide body temperature variation. Behavioural adaptations allow endothermic animals to exert some control over a given parameter. For instance, reptiles often rest on sun-heated rocks in the morning to raise their body temperature. Regulators are also responsive to external circumstances, however:

if the same sun-baked boulder happens to host a ground squirrel, the animal's metabolism will adjust to the lesser need for internal heat production. An advantage of homeostatic regulation is that it allows an organism to function effectively in a broad range of environmental conditions. For example, ectotherms tend to become sluggish at low temperatures, whereas a co-located endotherm may be fully active. That thermal stability comes at a price since an automatic regulation system requires additional energy. One reason snakes may eat only once a week is that they use much less energy to maintain homeostasis.

Most homeostatic regulation is controlled by the release of hormones into the bloodstream. However, other regulatory processes rely on simple diffusion to maintain a balance. Homeostatic regulation extends far beyond the control of temperature. All animals also regulate their blood glucose, as well as the concentration of their blood. Mammals regulate their blood glucose with insulin and glucagon. The human body maintains glucose levels constant most of the day, even after a 24-hour fast. Even during long periods of fasting, glucose levels are reduced only very slightly. Insulin, secreted by the beta cells of the pancreas, effectively transports glucose to the body's cells by instructing those cells to keep more of the glucose for their own use. seeDynamic equilibrium.

If the glucose inside the cells is high the cells will convert it to the insoluble glycogen to prevent the soluble glucose interfering with cellular metabolism. Ultimately this lowers blood glucose levels, and Insulin helps to prevent hyperglycemia. When insulin is deficient or cells become resistant to it, diabetes occurs. Glucagon, secreted by the alpha cells of the pancreas, encourages cells to break down stored glycogen or convert non-carbohydrate carbon sources to glucose via gluconeogenesis, thus preventing hypoglycemia. The kidneys are used to remove excess water and ions from the blood. These are then expelled as urine. The kidneys perform a vital role in homeostatic regulation in mammals, removing excess water, salt, and urea from the blood. These are the body's main waste products.

Another homeostatic regulation occurs in the gut. Homeostasis of the gut is not fully understood but it is believed that Toll-like receptor (TLR) expression profiles contribute to it. Intestinal epithelial cells exhibit important factors that contribute to homeostasis:1) They have different cellular distribution of TLR's compared to the normal gut mucosa. An example of this is how TLR5 (activated by flagellin) can redistribute to the basolateral membrane, which is the perfect place where flagellin can be detected. 2) The enterocytes express high levels of TLR inhibitor Toll-interacting protein (TOLLIP).

TOLLIP is a human gene that is a part of the innate immune system and is highest in a healthy gut; it correlates to luminal bacterial load. 3) Surface enterocytes also express high levels of Interleukin-1 receptor (IL-1R)-containing inhibitory molecule. IL-1R are also referred to as single immunoglobulin IL-

1R (SIGIRR). Animals deficient in this are more susceptible to induced colitis, implying that SIGIRR might possibly play a role in tuning mucosal tolerance towards commensal flora.

Nucleotide-binding oligomerisation domain containing 2 (NOD2) is suggested to have an effect on suppressing inflammatory cascades based on recent evidence. It is believed to modulate signals transmitted through TLRs, TLR3, 4, and 9 specifically. Mutation of it has resulted in Crohn's disease. Excessive T-helper 1 responses to resident flora in the gut are controlled by inhibiting the controlling influence of regulatory T cells and tolerance-inducing dendritic cells.

Sleep timing depends upon a balance between homeostatic sleep propensity, the need for sleep as a function of the amount of time elapsed since the last adequate sleep episode, and circadian rhythms that determine the ideal timing of a correctly structured and restorative sleep episode.

Control Mechanisms

All homeostatic control mechanisms have at least three interdependent components for the variable being regulated: The receptor is the sensing component that monitors and responds to changes in the environment. When the receptor senses a stimulus, it sends information to a control centre, the component that sets the range at which a variable is maintained. The control centre determines an appropriate response to the stimulus. In most homeostatic mechanisms the control centre is the brain. The control centre then sends signals to an effector, which can be muscles, organs or other structures that receive signals from the control centre. After receiving the signal, a change occurs to correct the deviation by either enhancing it with positive feedback or depressing it with negative feedback

Positive Feedback

Positive feedback mechanisms are designed to accelerate or enhance the output created by a stimulus that has already been activated.

Unlike negative feedback mechanisms that initiate to maintain or regulate physiological functions within a set and narrow range, the positive feedback mechanisms are designed to push levels out of normal ranges. To achieve this purpose, a series of events initiates a cascading process that builds to increase the effect of the stimulus. This process can be beneficial but is rarely used by the body due to risks of the acceleration's becoming uncontrollable.

One positive feedback example event in the body is blood platelet accumulation, which, in turn, causes blood clotting in response to a break or tear in the lining of blood vessels. Another example is the release of oxytocin to intensify the contractions that take place during childbirth.

Negative Feedback

Negative feedback mechanisms consist of reducing the output or activity of any organ or system back to its normal range of functioning. A good example of this is regulating blood pressure. Blood vessels can sense resistance of blood flow against the walls when blood pressure increases. The blood vessels act as the receptors and they relay this message to the brain.

The brain then sends a message to the heart and blood vessels, both of which are the effectors. The heart rate would decrease as the blood vessels increase in diameter (or vasodilation). This change would cause the blood pressure to fall back to its normal range. The opposite would happen when blood pressure decreases, and would cause vasoconstriction.

Another important example is seen when the body is deprived of food. The body would then reset the metabolic set point to a lower than normal value. This would allow the body to continue to function, at a slower rate, even though the body is starving. Therefore, people who deprive themselves of food while trying to lose weight would find it easy to shed weight initially and much harder to lose more after. This is due to the body readjusting itself to a lower metabolic set point to allow the body to survive with its low supply of energy. Exercise can change this effect by increasing the metabolic demand.

Another good example of negative feedback mechanism is temperature control. The hypothalamus, which monitors the body temperature, is capable of determining even the slightest of variation of normal body temperature (37 degrees Celsius). Response to such variation could be stimulation of glands that produces sweat to reduce the temperature or signalling various muscles to shiver to increase body temperature.

Both feedbacks are equally important for the healthy functioning of one's body. Complications can arise if any of the two feedbacks are affected or altered in any way.

Homeostatic Imbalance

Many diseases are a result of disturbance of homeostasis, a condition known as homeostatic imbalance. As it ages, every organism will lose efficiency in its control systems. The inefficiencies gradually result in an unstable internal environment that increases the risk for illness.

In addition, homeostatic imbalance is also responsible for the physical changes associated with aging. Even more serious than illness and other characteristics of aging is death. Heart failure has been seen where nominal negative feedback mechanisms become overwhelmed, and destructive positive feedback mechanisms then take over. Diseases that result from a homeostatic imbalance include diabetes, dehydration, hypoglycemia, hyperglycemia, gout, and any disease caused by a toxin present in the bloodstream. All of these conditions result from the presence of an increased amount of a particular substance. In ideal circumstances, homeostatic control mechanisms should prevent this imbalance from occurring, but, in some people, the mechanisms do not work efficiently enough or the quantity of the substance exceeds the levels at which it can be managed. In these cases, medical intervention is necessary to restore the balance, or permanent damage to the organs may result.

Varieties

The Dynamic Energy Budget theory for metabolic organization delineates structure and (one or more) reserves in an organism. Its formulation is based on three forms of homeostasis:

- Strong homeostasis, whereas structure and reserve do not change in composition. Because the amount of reserve and structure can vary, this allows a particular change in the composition of the whole body (as explained by the Dynamic Energy Budget theory).
- Weak homeostasis, wherein the ratio of the amounts of reserve and structure becomes constant as long as food availability is constant, even when the organism grows. This means that the whole body composition is constant during growth in constant environments.
- Structural homeostasis, wherein the sub-individual structures grow in harmony with the whole individual; the relative proportions of the individuals remain constant.

Ecological

Historically, ecological succession was seen as having a stable end-stage called the climax, sometimes referred to as the 'potential biodiversity' of a site, shaped primarily by the local climate. This idea has been largely abandoned by modern ecologists in favour of nonequilibrium ideas of how ecosystems function, as most natural ecosystems experience disturbance at a rate that makes a "climax" community unattainable.

Only on small, isolated habitats known as ecological islands can the phenomenon be observed. One such case study is the island of Krakatoa after its major eruption in 1883: the established stable homeostasis of the previous forest climax ecosystem was destroyed, and all life was eliminated from the island.

In the years after the eruption, Krakatoa went through a sequence of ecological changes in which successive groups of new plant or animal species followed one another, leading to increasing biodiversity and eventually culminating in a re-established climax community.

This *ecological succession* on Krakatoa occurred in a number of stages; a *sere* is defined as "a stage in a sequence of events by which succession occurs". The complete chain of seres leading to a climax is called a *prisere*.

In the case of Krakatoa, the island reached its climax community, with eight hundred different recorded species, in 1983, one hundred years after the eruption that cleared all life off the island. Evidence confirms that this number has been homeostatic for some time, with the introduction of new species rapidly leading to elimination of old ones. The evidence of Krakatoa, and other disturbed island ecosystems, has confirmed many principles of Island Biogeography, mimicking general principles of ecological succession albeit in a virtually closed system comprised almost exclusively of endemic species.

Biosphere

In the Gaia hypothesis, James Lovelock stated that the entire mass of living matter on Earth (or any planet with life) functions as a vast homeostatic superorganism that actively modifies its planetary environment to produce the environmental conditions necessary for its own survival.

Whether this sort of system is present on Earth is still open to debate. However, some relatively simple homeostatic mechanisms are generally accepted. For example, when atmospheric carbon dioxide levels rise, certain plants are able to grow better and thus act to remove more carbon dioxide from the atmosphere. When sunlight is plentiful and atmospheric temperature climbs, the phytoplankton of the ocean surface waters thrive and produce more dimethyl sulfide, DMS. The DMS molecules act as cloud condensation nuclei, which produce more clouds, and thus increase the atmospheric albedo, and this feeds back to lower the temperature of the atmosphere.

As scientists discover more about Earth, vast numbers of positive and negative feedback loops are being discovered, that, together, maintain a metastable condition, sometimes within very broad range of environmental conditions. Environmental pressure, such as competition or change in temperature, can lead to adaptation/extinction of species over time.

Reactive

Example of use: "Reactive homeostasis is an immediate homeostasic response to a challenge such as predation." However, *any* homeostasis is impossible without reaction-because homeostasis is and must be a "feedback" phenomenon. The phrase "reactive homeostasis" is simply short for "reactive compensation reestablishing homeostasis", that is to say, "reestablishing a point of homeostasis."-it should not be confused with a separate *kind* of homeostasis or a distinct phenomenon *from* homeostasis; it is simply the compensation (or compensatory) phase of homeostasis.

Glycolysis and Fermentation

Glycolysis (from *glycose*, an older term for glucose + *-lysis* degradation) is the metabolic pathway that converts glucose $C_6H_{12}O_6$, into pyruvate, $CH_3COCOO^- + H^+$. The free energy released in this process is used to form the high-energy compounds ATP (adenosine triphosphate) and NADH (reduced nicotinamide adenine dinucleotide). Glycolysis is a definite sequence of ten reactions involving ten intermediate compounds (one of the steps involves two intermediates).

The intermediates provide entry points to glycolysis. For example, most monosaccharides, such as fructose, glucose, and galactose, can be converted to one of these intermediates. The intermediates may also be directly useful. For example, the intermediate dihydroxyacetone phosphate is a source of the glycerol that combines with fatty acids to form fat. Glycolysis is thought to be the archetype of a universal metabolic pathway. It occurs, with variations, in nearly all organisms, both aerobic and anaerobic. The wide occurrence of glycolysis indicates that it is one of the most ancient known metabolic pathways.

The most common type of glycolysis is the *Embden-Meyerhof-Parnas pathway (EMP pathway)*, which was first discovered by Gustav Embden, Otto Meyerhof and Jakub Karol Parnas. Glycolysis also refers to other pathways, such as the *Entner–Doudoroff pathway* and various heterofermentative and homofermentative pathways. Dietary carbohydrate from which humans gain energy enter the body in complex forms, such as disaccharides and the polymers starch (amylose and amylopectin) and glycogen. The polymer cellulose is also consumed but not digested. The first step in the metabolism of digestible carbohydrate is the conversion of the higher polymers to simpler, soluble forms that can be transported across the intestinal wall and delivered to the tissues.

The breakdown of polymeric sugars begins in the mouth. Saliva has a slightly acidic pH of 6.8 and contains lingual amylase that begins the digestion of carbohydrates. The action of lingual amylase is limited to the area of the mouth and the esophagus; it is virtually inactivated by the much stronger acid pH of the stomach. Once the food has arrived in the stomach, acid hydrolysis contributes to its degradation; specific gastric proteases and lipases aid this process for proteins and fats, respectively. The mixture of gastric secretions, saliva, and food, known collectively as chyme, moves to the small intestine.

The main polymeric-carbohydrate digesting enzyme of the small intestine is α-amylase. This enzyme is secreted by the pancreas and has the same activity as salivary amylase, producing disaccharides and trisaccharides. The latter are converted to monosaccharides by intestinal saccharidases, including maltases that hydrolyze di- and trisaccharides, and the more specific disaccharidases, sucrase, lactase, and trehalase. The net result is the almost complete conversion of digestible carbohydrate to its constituent monosaccharides. The resultant glucose and other simple carbohydrates are transported across the intestinal wall to the hepatic portal vein and then to liver parenchymal cells and other tissues. There they are converted to fatty acids, amino acids, and glycogen, or else oxidized by the various catabolic pathways of cells.

Oxidation of glucose is known as glycolysis. Glucose is oxidized to either lactate or pyruvate. Under aerobic conditions, the dominant product in most tissues is pyruvate and the pathway is known as aerobic glycolysis. When oxygen is depleted, as for instance during prolonged vigorous exercise, the dominant glycolytic product in many tissues is lactate and the process is known as anaerobic glycolysis.

The Energy Derived from Glucose Oxidation

Aerobic glycolysis of glucose to pyruvate, requires two equivalents of ATP to activate the process, with the subsequent production of four equivalents of ATP and two equivalents of NADH. Thus, conversion of one mole of glucose to two moles of pyruvate is accompanied by the net production of two moles each of ATP and NADH.

$$\text{Glucose} + 2\ \text{ADP} + 2\ \text{NAD}^+ + 2\ \text{P}_i \rightarrow$$
$$2\ \text{Pyruvate} + 2\ \text{ATP} + 2\ \text{NADH} + 2\ \text{H}^+$$

The NADH generated during glycolysis is used to fuel mitochondrial ATP synthesis via oxidative phosphorylation, producing either two or three equivalents of ATP depending upon whether the glycerol phosphate shuttle or the malate-aspartate shuttle is used to transport the electrons from cytoplasmic NADH into the mitochondria.

The malate-aspartate shuttle is the principal mechanism for the movement of reducing equivalents (in the form of NADH) from the cytoplasm to the mitochondria. The glycolytic pathway is a primary source of NADH. Within the mitochodria the electrons of NADH can be coupled to ATP production during the process of oxidative phosphorylation.

The electrons are "carried" into the mitochondria in the form of malate. Cytoplasmic malate dehydrogenase (MDH) reduces oxaloacetate (OAA) to malate while oxidizing NADH to NAD^+. Malate then enters the mitochondria where the reverse reaction is carried out by mitochondrial MDH. Movement of mitochondrial OAA to the cytoplasm to maintain this cycle requires it be transaminated to aspartate (Asp, D) with the amino group being donated by glutamate (Glu, E). The Asp then leaves the mitochondria and enters the cytoplasm. The deamination of glutamate generates α-ketoglutarate (α-KG) which leaves the mitochondria for the cytoplasm.

All the participants in the cycle are present in the proper cellular compartment for the shuttle to function due to concentration dependent movement. When the energy level of the cell rises the rate of mitochondrial oxidation of NADH to NAD^+ declines and therefore, the shuttle slows. G3PDH is glyceraldehyde-3-phosphate dehydrogenase.

The glycerol phosphate shuttle is a secondary mechanism for the transport of electrons from cytosolic NADH to mitochondrial carriers of the oxidative phosphorylation pathway. The primary cytoplasmic NADH electron shuttle is the malate-aspartate shuttle.

Two enzymes are involved in this shuttle. One is the cytosolic version of the enzyme glycerol-3-phosphate dehydrogenase (glycerol-3-PDH) which has as one substrate, NADH. The second is the mitochondrial form of the enzyme which has as one of its' substrates, FAD^+. The net result is that there is a continual conversion of the glycolytic intermediate, DHAP and glycerol-3-phosphate with the concomitant transfer of the electrons from reduced cytosolic NADH to mitochondrial oxidized FAD^+. Since the electrons from mitochondrial $FADH_2$ feed into the oxidative phosphorylation pathway at coenzyme Q (as opposed to NADH-ubiquinone oxidoreductase [complex I]) only 2 moles of ATP will be generated from glycolysis. G3PDH is glyceraldehyde-3-phoshate dehydrogenase. The net yield from the oxidation of 1 mole of glucose to 2 moles of pyruvate is, therefore, either 6 or 8 moles of ATP. Complete oxidation of the 2 moles of pyruvate, through the TCA cycle, yields an additional 30 moles of ATP; the total yield, therefore being either 36 or 38 moles of ATP from the complete oxidation of 1 mole of glucose to CO_2 and H_2O.

The Individual Reactions of Glycolysis

The pathway of glycolysis can be seen as consisting of 2 separate phases. The first is the chemical priming phase requiring energy in the form of ATP, and the second is considered the energy-yielding phase. In the first phase, 2 equivalents of ATP are used to convert glucose to fructose 1,6-bisphosphate (F1,6BP). In the second phase F1,6BP is degraded to pyruvate, with the production of 4 equivalents of ATP and 2 equivalents of NADH.

Pathway of glycolysis from glucose to pyruvate. Substrates and products are in blue, enzymes are in green. The two high energy intermediates whose oxidations are coupled to ATP synthesis are shown in red (1,3-bisphosphoglycerate and phosphoenolpyruvate).

The Hexokinase Reaction

The ATP-dependent phosphorylation of glucose to form glucose 6-phosphate (G6P) is the first reaction of glycolysis, and is catalysed by tissue-specific isoenzymes known as hexokinases. The phosphorylation accomplishes two goals: First, the hexokinase reaction converts nonionic glucose into an anion that is trapped in the cell, since cells lack transport systems for phosphorylated sugars. Second, the otherwise biologically inert glucose becomes activated into a labile form capable of being further metabolized.

Four mammalian isozymes of hexokinase are known (Types I–IV), with the Type IV isozyme often referred to as glucokinase. Glucokinase is the form of the enzyme found in hepatocytes and pancreatic β-cells. The high K_m of glucokinase for glucose means that this enzyme is saturated only at very high concentrations of substrate.

This feature of hepatic glucokinase allows the liver to buffer blood glucose. After meals, when postprandial blood glucose levels are high, liver glucokinase is significantly active, which causes the liver preferentially to trap and to store circulating glucose.

When blood glucose falls to very low levels, tissues such as liver and kidney, which contain glucokinases but are not highly dependent on glucose, do not continue to use the meager glucose supplies that remain available.

At the same time, tissues such as the brain, which are critically dependent on glucose, continue to scavenge blood glucose using their low K_m hexokinases, and as a consequence their viability is protected. Under various conditions of glucose deficiency, such as long periods between meals, the liver is stimulated to supply the blood with glucose through the pathway of gluconeogenesis.

The levels of glucose produced during gluconeogenesis are insufficient to activate glucokinase, allowing the glucose to pass out of hepatocytes and into the blood.

The regulation of hexokinase and glucokinase activities is also different. Hexokinases I, II, and III are allosterically inhibited by product (G6P) accumulation, whereas glucokinases are not. The latter further insures liver accumulation of glucose stores during times of glucose excess, while favouring peripheral glucose utilization when glucose is required to supply energy to peripheral tissues.

Phosphohexose Isomerase

The second reaction of glycolysis is an isomerization, in which G6P is converted to fructose 6-phosphate (F6P). The enzyme catalysing this reaction is phosphohexose isomerase. The reaction is freely reversible at normal cellular concentrations of the two hexose phosphates and thus catalyzes this interconversion during glycolytic carbon flow and during gluconeogenesis.

6-Phosphofructo-1-Kinase (Phosphofructokinase-1, PFK-1): The next reaction of glycolysis involves the utilization of a second ATP to convert F6P to fructose 1,6-bisphosphate (F1,6BP). This reaction is catalysed by 6-

phosphofructo-1-kinase, better known as phosphofructokinase-1 or PFK-1. This reaction is not readily reversible because of its large positive free energy ($\Delta G^{0'}$ = +5.4 kcal/mol) in the reverse direction. Nevertheless, fructose units readily flow in the reverse (gluconeogenic) direction because of the ubiquitous presence of the hydrolytic enzyme, fructose-1,6-bisphosphatase (F-1,6-BPase).

The presence of these two enzymes in the same cell compartment provides an example of a metabolic futile cycle, which if unregulated would rapidly deplete cell energy stores. However, the activity of these two enzymes is so highly regulated that PFK-1 is considered to be the rate-limiting enzyme of glycolysis and F-1,6-BPase is considered to be the rate-limiting enzyme in gluconeogenesis.

Aldolase

Aldolase catalyses the hydrolysis of F1,6BP into two 3-carbon products: dihydroxyacetone phosphate (DHAP) and glyceraldehyde 3-phosphate (G3P). The aldolase reaction proceeds readily in the reverse direction, being utilized for both glycolysis and gluconeogenesis.

Triose Phosphate Isomerase

The two products of the aldolase reaction equilibrate readily in a reaction catalysed by triose phosphate isomerase. Succeeding reactions of glycolysis utilize G3P as a substrate; thus, the aldolase reaction is pulled in the glycolytic direction by mass action principals.

Glyceraldehyde-3-Phosphate Dehydrogenase

The second phase of glucose catabolism features the energy-yielding glycolytic reactions that produce ATP and NADH. In the first of these reactions, glyceraldehyde-3-P dehydrogenase (G3PDH) catalyzes the NAD^+-dependent oxidation of G3P to 1,3-bisphosphoglycerate (1,3BPG) and NADH. The G3PDH reaction is reversible, and the same enzyme catalyzes the reverse reaction during gluconeogenesis.

Phosphoglycerate Kinase

The high-energy phosphate of 1,3-BPG is used to form ATP and 3-phosphoglycerate (3PG) by the enzyme phosphoglycerate kinase. Note that this is the only reaction of glycolysis or gluconeogenesis that involves ATP and yet is reversible under normal cell conditions. Associated with the phosphoglycerate kinase pathway is an important reaction of erythrocytes, the formation of 2,3-bisphosphoglycerate, 2,3BPG by the enzyme bisphosphoglycerate mutase. 2,3BPG is an important regulator of hemoglobin's affinity for oxygen. Note that 2,3-bisphosphoglycerate phosphatase degrades 2,3BPG to 3-phosphoglycerate, a normal intermediate of glycolysis. The 2,3BPG shunt thus operates with the expenditure of 1 equivalent of ATP per triose passed through the shunt. The process is not reversible under physiological conditions. The pathway for 2,3-bisphosphoglycerate (2,3-BPG) synthesis within erythrocytes. Synthesis of 2,3-BPG represents a major reaction pathway for the consumption of glucose in erythrocytes. The synthesis of 2,3-BPG in erythrocytes is critical for controlling hemoglobin affinity for oxygen. Note that when glucose is oxidized by this pathway the erythrocyte loses the ability to gain 2 moles of ATP from glycolytic oxidation of 1,3-BPG to 3-phosphoglycerate via the phosphoglycerate kinase reaction.

Phosphoglycerate Mutase and Enolase

The remaining reactions of glycolysis are aimed at converting the relatively low energy phosphoacyl-ester of 3PG to a high-energy form and harvesting the phosphate as ATP. The 3PG is first converted to 2PG by phosphoglycerate mutase and the 2PG conversion to phosphoenoylpyruvate (PEP) is catalysed by enolase.

Pyruvate Kinase

The final reaction of aerobic glycolysis is catalysed by the highly regulated enzyme pyruvate kinase (PK). In this strongly exergonic reaction, the high-energy phosphate of PEP is conserved as ATP. The loss of phosphate by PEP leads to the production of pyruvate in an unstable enol form, which spontaneously tautomerizes to the more stable, keto form of pyruvate.

This reaction contributes a large proportion of the free energy of hydrolysis of PEP. There are two distinct genes encoding PK activity. One is located on chromosome 1 and encodes the liver and erythrocyte PK proteins (identified as the PKLR gene) and the other is located on chromosome 15 and encodes the muscle PK proteins (identified as the PKM gene). The muscle PKM gene directs the synthesis of two isoforms of muscle PK termed PK-M1 and PK-M2. Deficiencies in the PKLR gene are the cause of the most common form of inherited non-spherocytic anemia.

Anaerobic Glycolysis

Under aerobic conditions, pyruvate in most cells is further metabolized via the TCA cycle. Under anaerobic conditions and in erythrocytes under aerobic conditions, pyruvate is converted to lactate by the enzyme lactate dehydrogenase (LDH), and the lactate is transported out of the cell into the circulation. The conversion of pyruvate to lactate, under anaerobic conditions, provides the cell with a mechanism for the oxidation of NADH (produced during the G3PDH reaction) to NAD^+ which occurs during the LDH catalysed reaction. This reduction is required since NAD^+ is a necessary substrate for G3PDH, without which glycolysis will cease. Normally, during aerobic glycolysis the electrons of cytoplasmic NADH are transferred to mitochondrial carriers of the oxidative phosphorylation pathway generating a continuous pool of cytoplasmic NAD^+. Aerobic glycolysis generates substantially more ATP per mole of glucose oxidized than does anaerobic glycolysis.

The utility of anaerobic glycolysis, to a muscle cell when it needs large amounts of energy, stems from the fact that the rate of ATP production from glycolysis is approximately 100X faster than from oxidative phosphorylation. During exertion muscle cells do not need to energize anabolic reaction pathways. The requirement is to generate the maximum amount of ATP, for muscle contraction, in the shortest time frame. This is why muscle cells derive almost all of the ATP consumed during exertion from anaerobic glycolysis.

Regulation of Glycolysis

The reactions catalysed by hexokinase, PFK-1 and PK all proceed with a relatively large free energy decrease. These non-equilibrium reactions of glycolysis would be ideal candidates for regulation of the flux through glycolysis. Indeed, *in vitro* studies have shown all three enzymes to be allosterically controlled.

Regulation of hexokinase, however, is not the major control point in glycolysis. This is due to the fact that large amounts of G6P are derived from the breakdown of glycogen (the predominant mechanism of carbohydrate entry into glycolysis in skeletal muscle) and, therefore, the hexokinase reaction is not necessary. Regulation of PK is important for reversing glycolysis when ATP is high in order to activate gluconeogenesis. As such this enzyme catalysed reaction is not a major control point in glycolysis. The rate limiting step in glycolysis is the reaction catalysed by PFK-1.

PFK-1 is a tetrameric enzyme that exist in two conformational states termed R and T that are in equilibrium. ATP is both a substrate and an allosteric inhibitor of PFK-1. Each subunit has two ATP binding sites, a substrate site and an inhibitor site. The substrate site binds ATP equally well when the tetramer is in either conformation.

The inhibitor site binds ATP essentially only when the enzyme is in the T state. F6P is the other substrate for PFK-1 and it also binds preferentially to the R state enzyme. At high concentrations of ATP, the inhibitor site becomes occupied and shifting the equilibrium of PFK-1 conformation to that of the T state decreasing PFK-1's ability to bind F6P.

The inhibition of PFK-1 by ATP is overcome by AMP which binds to the R state of the enzyme and, therefore, stabilizes the conformation of the enzyme capable of binding F6P. The most important allosteric regulator of both glycolysis and gluconeogenesis is fructose 2,6-bisphosphate, F2,6BP, which is not an intermediate in glycolysis or in gluconeogenesis.

Regulation of glycolysis and gluconeogenesis by fructose 2,6-bisphosphate (F2,6BP). The major sites for regulation of glycolysis and gluconeogenesis are the phosphofructokinase-1 (PFK-1) and fructose-1,6-bisphosphatase (F-1,6-BPase) catalysed reactions. PFK-2 is the kinase activity and F-2,6-BPase is the phosphatase activity of the bi-functional regulatory enzyme, phosphofructokinase-2/fructose-2,6-bisphosphatase. PKA is cAMP-dependent protein kinase which phosphorylates PFK-2/F-2,6-BPase turning on the phosphatase activity. (+ve) and (-ve) refer to positive and negative activities, respectively.

The synthesis of F2,6BP is catalysed by the bifunctional enzyme phosphofructokinase-2/fructose-2,6-bisphosphatase (PFK-2/F-2,6-BPase). In the nonphosphorylated form the enzyme is known as PFK-2 and serves to catalyse the synthesis of F2,6BP by phosphorylating fructose 6-phosphate. The result is that the activity of PFK-1 is greatly stimulated and the activity of F-1,6-BPase is greatly inhibited.

Under conditions where PFK-2 is active, fructose flow through the PFK-1/F-1,6-BPase reactions takes place in the glycolytic direction, with a net production of F1,6BP. When the bifunctional enzyme is phosphorylated it no longer exhibits kinase activity, but a new active site hydrolyzes F2,6BP to F6P and inorganic phosphate. The metabolic result of the phosphorylation of the bifunctional enzyme is that allosteric stimulation of PFK-1 ceases, allosteric inhibition of F-1,6-BPase is eliminated, and net flow of fructose through these two enzymes is gluconeogenic, producing F6P and eventually glucose. The interconversion of the bifunctional enzyme is catalysed by cAMP-dependent protein kinase (PKA), which in turn is regulated by circulating peptide hormones.

When blood glucose levels drop, pancreatic insulin production falls, glucagon secretion is stimulated, and circulating glucagon is highly increased. Hormones such as glucagon bind to plasma membrane receptors on liver cells, activating membrane-localized adenylate cyclase leading to an increase in the conversion of ATP to cAMP. cAMP binds to the regulatory subunits of PKA, leading to release and activation of the catalytic subunits.

PKA phosphorylates numerous enzymes, including the bifunctional PFK-2/F-2,6-BPase. Under these conditions the liver stops consuming glucose and becomes metabolically gluconeogenic, producing glucose to reestablish normoglycemia.

Representative pathway for the activation of cAMP-dependent protein kinase (PKA). In this example glucagon binds to its' cell-surface receptor, thereby activating the receptor. Activation of the receptor is coupled to the activation of a receptor-coupled G-protein (GTP-binding and hydrolyzing protein) composed of 3 subunits. Upon activation the alpha subunit dissociates and binds to and activates adenylate cyclase. Adenylate cylcase then converts ATP to cyclic-AMP (cAMP). The cAMP thus produced then binds to the regulatory subunits of PKA leading to dissociation of the associated catalytic subunits. The catalytic subunits are inactive until dissociated from the regulatory subunits. Once released the catalytic subunits of PKA phosphorylate numerous substrate using ATP as the phosphate donor.

Regulation of glycolysis also occurs at the step catalysed by pyruvate kinase, (PK). The liver enzyme has been most studied *in vitro*. This enzyme is inhibited by ATP and acetyl-CoA and is activated by F1,6BP. The inhibition of PK by ATP is similar to the effect of ATP on PFK-1. The binding of ATP to the inhibitor site reduces its affinity for PEP. The liver enzyme is also controlled at the level of synthesis. Increased carbohydrate ingestion induces the synthesis of PK resulting in elevated cellular levels of the enzyme.

A number of PK isozymes have been described. The liver isozyme (L-type), characteristic of a gluconeogenic tissue, is regulated via phosphorylation by PKA, whereas the M-type isozyme found in brain, muscle, and other glucose requiring tissue is unaffected by PKA. As a consequence of these differences, blood glucose levels and associated hormones can regulate the balance of liver gluconeogenesis and glycolysis while muscle metabolism remains unaffected.

In erythrocytes, the fetal PK isozyme has much greater activity than the adult isozyme; as a result, fetal erythrocytes have comparatively low concentrations of glycolytic intermediates. Because of the low steady-state concentration of fetal 1,3BPG, the 2,3BPG shunt is greatly reduced in fetal cells and little 2,3BPG is formed. Since 2,3BPG is a negative effector of hemoglobin affinity for oxygen, fetal erythrocytes have a higher oxygen affinity than maternal erythrocytes. Therefore, transfer of oxygen from maternal hemoglobin to fetal hemoglobin is favoured, assuring the fetal oxygen supply. In the newborn, an erythrocyte isozyme of the M-type with comparatively low PK activity displaces the fetal type, resulting in an accumulation of glycolytic intermediates. The increased 1,3BPG levels activate the 2,3BPG shunt, producing 2,3BPG needed to regulate oxygen binding to hemoglobin.

Genetic diseases of adult erythrocyte PK are known in which the kinase is virtually inactive. The erythrocytes of affected individuals have a greatly reduced capacity to make ATP and thus do not have sufficient ATP to perform activities such as ion pumping and maintaining osmotic balance. These erythrocytes have a short half-life, lyse readily, and are responsible for some cases of hereditary hemolytic anemia.

The liver PK isozyme is regulated by phosphorylation, allosteric effectors, and modulation of gene expression. The major allosteric effectors are F1,6BP, which stimulates PK activity by decreasing its K_m for PEP, and for the negative effector, ATP. Expression of the liver PK gene is strongly influenced by the quantity of carbohydrate in the diet, with high-carbohydrate diets inducing up to 10-fold increases in PK concentration as compared to low carbohydrate diets. Liver PK is phosphorylated and inhibited by PKA, and thus it is under hormonal control similar to that described earlier for PFK-2.

Muscle PK (M-type) is not regulated by the same mechanisms as the liver enzyme. Extracellular conditions that lead to the phosphorylation and inhibition of liver PK, such as low blood glucose and high levels of circulating glucagon, do not inhibit the muscle enzyme. The result of this differential regulation is that hormones such as glucagon and epinephrine favour liver gluconeogenesis by inhibiting liver glycolysis, while at the same time, muscle glycolysis can proceed in accord with needs directed by intracellular conditions.

Metabolic Fates of Pyruvate

Pyruvate is the branch point molecule of glycolysis. The ultimate fate of pyruvate depends on the oxidation state of the cell. In the reaction catalysed by G3PDH a molecule of NAD^+ is reduced to NADH. In order to maintain the re-dox state of the cell, this NADH must be re-oxidized to NAD^+. During aerobic glycolysis this occurs in the mitochondrial electron transport chain generating ATP. Thus, during aerobic glycolysis ATP is generated from oxidation of glucose directly at the PGK and PK reactions as well as indirectly by re-oxidation of NADH in the oxidative phosphorylation pathway. Additional NADH molecules are generated during the complete aerobic oxidation of pyruvate in the TCA cycle. Pyruvate enters the TCA cycle in the form of acetyl-CoA which is the product of the pyruvate dehydrogenase reaction. The fate of pyruvate during anaerobic glycolysis is reduction to lactate.

Lactate Metabolism

During anaerobic glycolysis, that period of time when glycolysis is proceeding at a high rate (or in anaerobic organisms), the oxidation of NADH occurs through the reduction of an organic substrate. Erythrocytes and skeletal muscle (under conditions of exertion) derive all of their ATP needs through anaerobic glycolysis. The large quantity of NADH produced is oxidized by reducing pyruvate to lactate. This reaction is carried out by lactate dehydrogenase, (LDH). The lactate produced during anaerobic glycolysis diffuses from the tissues and is transported to highly aerobic tissues such as cardiac muscle and liver. The lactate is then oxidized to pyruvate in these cells by LDH and the pyruvate is further oxidized in the TCA cycle. If the energy level in these cells is high the carbons of pyruvate will be diverted back to glucose via the gluconeogenesis pathway.

Mammalian cells contain two distinct types of LDH subunits, termed M and H. Combinations of these different subunits generates LDH isozymes with different characteristics. The H type subunit predominates in aerobic tissues such as heart muscle (as the H4 tetramer) while the M subunit predominates in anaerobic tissues such as skeletal muscle as the M4 tetramer). H4 LDH has a low K_m for pyruvate and also is inhibited by high levels of pyruvate. The M4 LDH enzyme has a high K_m for pyruvate and is not inhibited by pyruvate. This suggests that the H-type LDH is utilized for oxidizing lactate to pyruvate and the M-type the reverse.

Ethanol Metabolism

Metabolism of ethanol occurs primarily in the liver and involves at least three enzymatic pathways. The most significant pathway, responsible for the bulk of ethanol metabolism, is that initiated by alcohol dehydrogenase, ADH. The second major pathway for ethanol metabolism is the microsomal ethanol oxidizing system (MEOS) which involves the cytochrome P450 enzyme CYP2E1 and requires NADPH instead of NAD^+ as for ADH. The MEOS pathway is induced in individuals who chronically consume alcohol. The third pathway involves a non-oxidative pathway catalysed by fatty acid ethyl ester (FAEE) synthase. This latter pathway results in the formation of fatty acid ethyl esters and takes place primarily in the liver and pancreas, both of which are highly susceptible to the toxic effects of alcohol. Animal cells (primarily hepatocytes) contain cytosolic ADH which oxidizes ethanol to acetaldehyde. Acetaldehyde then enters the mitochondria where it is oxidized to acetate by acetaldehyde dehydrogenase (AcDH).

In humans there are multiple isoforms of ADH encoded for by at least seven different ADH genes. Functional ADH exists as a dimer and can be classified by five distinct classes (denoted I–V). The hepatic form of ADH is class I and consists of A, B, and C subunits encoded by three different ADH genes (ADH1A, ADH1B, and ADH1C). It should be pointed out that humans evolved to express multiple ADH genes and isoforms, not for metabolism of ethanol, but to metabolize naturally occurring alcohols found in foods as well as those produced by intestinal bacteria. As an example, one form of ADH is responsible for the metabolism of retinol (vitamin A) to retinaldehyde which is the form of vitamin A necessary for vision.

Acetaldehyde forms adducts with proteins, nucleic acids and other compounds, the results of which are the toxic side effects (the hangover) that are associated with alcohol consumption. The ADH and AcDH catalysed reactions also leads to the reduction of NAD^+ to NADH. The metabolic effects of ethanol intoxication stem from the actions of ADH and AcDH and the resultant cellular imbalance in the $NADH/NAD^+$. The NADH produced in the cytosol by ADH must be reduced back to NAD^+ via either the malate-aspartate shuttle or the glycerol-phosphate shuttle. Thus, the ability of an individual to metabolize ethanol is dependent upon the capacity of hepatocytes to carry out either of these 2 shuttles, which in turn is affected by the rate of the TCA cycle in the mitochondria whose rate of function is being impacted by the NADH produced by the AcDH reaction.

The reduction in NAD^+ impairs the flux of glucose through glycolysis at the glyceraldehyde-3-phosphate dehydrogenase reaction, thereby limiting energy production. Additionally, there is an increased rate of hepatic lactate production due to the effect of increased NADH on direction of the hepatic lactate dehydrogenase (LDH) reaction. This reversal of the LDH reaction in hepatocytes diverts pyruvate from gluconeogenesis leading to a reduction in the capacity of the liver to deliver glucose to the blood.

In addition to the negative effects of the altered NADH/NAD^+ ratio on hepatic gluconeogenesis, fatty acid oxidation is also reduced as this process requires NAD^+ as a cofactor. In fact the opposite is true, fatty acid synthesis is increased and there is an increase in triacylglyceride production by the liver. In the mitochondria, the production of acetate from acetaldehyde leads to increased levels of acetyl-CoA. Since the increased generation of NADH also reduces the activity of the TCA cycle, the acetyl-CoA is diverted to fatty acid synthesis. The reduction in cytosolic NAD^+ leads to reduced activity of glycerol-3-phosphate dehydrogenase (in the glycerol 3-phosphate to DHAP direction) resulting in increased levels of glycerol 3-phosphate which is the backbone for the synthesis of the triacylglycerides. Both of these two events lead to fatty acid deposition in the liver leading to fatty liver syndrome.

The MEOS pathway for ethanol metabolism is induced in chronic alcoholics and may contribute to the tolerance to ethanol observed in alcoholics. The activity of CYP2E1 is also essential in the metabolism of several xenobiotics. Therefore, the increased level of expression of this enzyme in alcoholics can have a significant impact on the production of toxic metabolites and this is thought to contribute to ethanol-induced liver injury. Of particular clinical significance is that increased levels of CYP2E1 result in accelerated conversion of the non-steroidal anti-inflammatory (NSAID) acetaminophen (Tylenol®) into toxic metabolites. Metabolism of ethanol by CYP2E1 also results in a significant increase in free radical and acetaldehyde production which, in turn, diminish reduced glutathione (GSH) and other defence systems against oxidative stress leading to further hepatocyte damage. Along with increased CYP2E1 activity there is an induction of microsomal enzymes involved in lipoprotein production, resulting in hyperlipemia which contributes to the fatty liver syndrome.

Regulation of Blood Glucose Levels

If for no other reason, it is because of the demands of the brain for oxidizable glucose that the human body exquisitely regulates the level of glucose circulating in the blood. This level is maintained in the range of 5mM. Nearly all carbohydrates ingested in the diet are converted to glucose following transport to the liver. Catabolism of dietary or cellular proteins generates carbon atoms that can be utilized for glucose synthesis via gluconeogenesis. Additionally, other tissues besides the liver that incompletely oxidize glucose (predominantly skeletal muscle and erythrocytes) provide lactate that can be converted to glucose via gluconeogenesis.

Maintenance of blood glucose homeostasis is of paramount importance to the survival of the human organism. The predominant tissue responding to signals that indicate reduced or elevated blood glucose levels is the liver. Indeed, one of the most important functions of the liver is to produce glucose for the circulation. Both elevated and reduced levels of blood glucose trigger hormonal responses to initiate pathways designed to restore glucose homeostasis.

Low blood glucose triggers release of glucagon from pancreatic α-cells. High blood glucose triggers release of insulin from pancreatic β-cells. Additional signals, ACTH and growth hormone, released from the pituitary act to increase blood glucose by inhibiting uptake by extrahepatic tissues. Glucocorticoids also act to increase blood glucose levels by inhibiting glucose uptake. Cortisol, the major glucocorticoid released from the adrenal cortex, is secreted in response to the increase in circulating ACTH. The adrenal medullary hormone, epinephrine, stimulates production of glucose by activating glycogenolysis in response to stressful stimuli.

Glucagon binding to its' receptors on the surface of liver cells triggers an increase in cAMP production leading to an increased rate of glycogenolysis by activating glycogen phosphorylase via the PKA-mediated cascade. This is the same response hepatocytes have to epinephrine release. The resultant increased levels of G6P in hepatocytes is hydrolysed to free glucose, by glucose-6-phosphatase, which then diffuses to the blood. The glucose enters extrahepatic cells where it is re-phosphorylated by hexokinase.

Since muscle and brain cells lack glucose-6-phosphatase, the glucose-6-phosphate product of hexokinase is retained and oxidized by these tissues. In opposition to the cellular responses to glucagon (and epinephrine on hepatocytes), insulin stimulates extrahepatic uptake of glucose from the blood and inhibits glycogenolysis in extrahepatic cells and conversely stimulates glycogen synthesis. As the glucose enters hepatocytes it binds to and inhibits glycogen phosphorylase activity.

The binding of free glucose stimulates the de-phosphorylation of phosphorylase thereby, inactivating it. Why is it that the glucose that enters hepatocytes is not immediately phosphorylated and oxidized? Liver cells contain an isoform of hexokinase called glucokinase. Glucokinase has a much lower affinity for glucose than does hexokinase. Therefore, it is not fully active at the physiological ranges of blood glucose. Additionally, glucokinase is not inhibited by its product G6P, whereas, hexokinase is inhibited by G6P.

Hepatocytes, unlike most other cells, are freely permeable to glucose and are, therefore, essentially unaffected by the action of insulin at the level of increased glucose uptake. When blood glucose levels are low, the liver does not compete with other tissues for glucose since the extrahepatic uptake of glucose is stimulated in response to insulin. Conversely, when blood glucose levels are high extrahepatic needs are satisfied and the liver takes up glucose for conversion into glycogen for future needs. Under conditions of high blood glucose, liver glucose levels will be high and the activity of glucokinase will be elevated. The G6P produced by glucokinase is rapidly converted to G1P by phosphoglucomutase, where it can then be incorporated into glycogen.

Role of the Kidney in Blood Glucose Control

Although the liver is the major site of glucose homeostasis, the kidney plays a vital role in the overall process of regulating the level of blood glucose. The kidney carries out gluconeogenesis primarily using the carbon skeleton of glutamine and while so doing allows for the elimination of waste nitrogen and maintaining plasma pH balance.

In addition to carrying out gluconeogenesis, the kidney regulates blood glucose levels via its ability to excrete glucose via glomerular filtration as well as to reabsorb the filtered glucose in the proximal convoluted tubules. The average adult will kidneys will filter around 180gm of glucose per day. Of this amount less than 1% is excreted in the urine due to efficient reabsorption. This reabsorption process is critical for maintaining blood glucose homeostasis and for retaining important calories for energy production.

Transport of glucose from the tubule into the tubular epithelial cells is carried out by specialized transport proteins termed sodium-glucose co-transporters (SGLTs). The SGLTs represent a family of transporters that are involved in the transport of glucose, amino acids, vitamins, and ions and other osmolytes across the brush-border membranes of kidney tubule cells and intestinal epithelial cells.

There are two SGLTs in the kidney involved in glucose reabsorption. SGLT1 is found primarily in the distal S2/S3 segment of the proximal tubule and SGLT2 is expressed exclusively in the S1 segment. The location of SGLT2 in the proximal tubule means that it is primarily responsible for glucose reabsorption. SGLT2 is a high-capacity low-affinity transporter that due to its expression location is responsible for approximately 90% of the glucose reabsorption activity of the kidney.

As would be expected from the name of the renal glucose transporters, SGLT1 and SGLT2 catalyse the active transport of glucose against a concentration gradient across the lumenal (apical) membrane of the tubule cell and couple this transport to sodium uptake. The inward sodium uptake is maintained by ATP-driven active transport of the sodium across the basolateral (anti-lumenal) membrane into the blood (coupled to inward uptake of potassium).

The reabsorbed glucose passively diffuses out of the tubule cell into the blood via the basolateral membrane associated GLUT2. Under normal conditions saturation of the ability of SGLT2 (and SGLT1) to reabsorb glucose is never saturated. The kidney can filter and reabsorb approximately 375mg of glucose per minute. The plasma concentration of glucose required to exceed this capacity is well above that considered normal and is only observed situations of renal dysfunction/disease or most importantly in type 2 diabetes.

Because of the importance of SGLT2 in renal reabsorption of glucose this transporter has become the target for therapeutic intervention of the hyperglycemia associated with type 2 diabetes. By specifically inhibiting SGLT2 there will be increased glucose excretion in the urine and thus a lowering of plasma glucose levels. Several SGLT2-specific inhibitors are currently in clinical trials with a few reaching phase III status.

Glucose Transporters

One major response of non-hepatic tissues to insulin is the recruitment, to the cell surface, of glucose transporter complexes. Glucose transporters comprise a family of at least 14 members. The most well characterized members of the family are GLUT1, GLUT2, GLUT3, GLUT4 and GLUT5. The glucose transporters are facilitative transporters that carry hexose sugars across the membrane without requiring energy. These transporters belong to a family of proteins called the solute carriers. Specifically, the official gene names for the GLUTs are solute carrier family 2 (facilitated glucose transporter) member. Thus, the GLUT1 gene symbol is SLC2A1, GLUT2 is SLC2A2, GLUT3 is SLC2A3, GLUT4 is SLC2A4 and GLUT5 is SLC2A5.

The glucose transporters can be divided into three classes based upon primary amino acid sequence comparisons. Class I transporters include GLUT1, GLUT2, GLUT3 (and the gene duplication of GLUT3 identified as GLUT14), and GLUT4. Class II transporters include GLUT5, GLUT7, GLUT9 and GLUT11. Class III transporters include GLUT6, GLUT8, GLUT10, GLUT12 and HMIT [proton (H^+) myoinositol symporter: SLC2A13]. HMIT is also known as GLUT13. GLUT1 is ubiquitously distributed in various tissues with highest levels of expression seen in erythrocytes. In fact in erythrocytes GLUT1 accounts for almost 5% of total protein. Although widely expressed GLUT1 is not expressed in hepatocytes.

GLUT2 is found primarily in intestine, pancreatic β-cells, kidney and liver. The K_m of GLUT2 for glucose (17mM) is the highest of all the sugar transporters. The high K_m ensures a fast equilibrium of glucose between the cytosol and the extracellular space ensuring that liver and pancreas do not metabolize glucose until its levels rise sufficiently in the blood. GLUT2 molecules can transport both glucose and fructose. When the concentration of blood glucose increases in response to food intake, pancreatic GLUT2 molecules mediate an increase in glucose uptake which leads to increased insulin secretion. For this reason, GLUT2 is thought to be a "glucose sensor".

GLUT3 is found primarily in neurons but also found in the intestine. GLUT3 binds glucose with high affinity (has the lowest K_m of the GLUTs) which allows neurons to have enhanced access to glucose especially under conditions of low blood glucose. Insulin-sensitive tissues, such as skeletal muscle and adipose tissue, contain GLUT4 whose mobilization to the cell-surface is stimulated by insulin action.

GLUT5 and the closely related transporter GLUT7 are involved in fructose transport. GLUT5 is expressed in intestine, kidney, testes, skeletal muscle, adipose tissue and brain. Although GLUT2, -5, -7, 8, -9, -11, and -12 can all transport fructose, GLUT5 is the only transporter that exclusively transports fructose.

GLUT9 (SLC2A9) does not transport sugar but is a uric acid transporter abundant in the kidney and liver.

Recent evidence has shown that the cell surface receptor for the human T cell leukemia virus (HTLV) is the ubiquitous GLUT1.

Fermentation

There are two important ways a cell can harvest energy from food: fermentation and cellular respiration. Both start with the same first step: the process of glycolysis which is the breakdown or splitting of glucose (6 carbons) into two 3-carbon molecules called pyruvic acid. The energy from other sugars, such as fructose, is also harvested using this process. Glycolysis is probably the oldest known way of producing ATP. There is evidence that the process of glycolysis predates the existence of O_2 in the Earth's atmosphere and organelles in cells:

- Glycolysis does not need oxygen as part of any of its chemical reactions. It serves as a first step in a variety of both aerobic and anaerobic energy-harvesting reactions.
- Glycolysis happens in the cytoplasm of cells, not in some specialized organelle.

Glycolysis is the one metabolic pathway found in all living organisms.

Fermenation

Fermentation is the process of deriving energy from the oxidation of organic compounds, such as carbohydrates, and using an endogenous electron acceptor, which is usually an organic compound. In contrast, respiration is where electrons are donated to an exogenous electron acceptor, such as oxygen, via an electron transport chain.

Fermentation is important in anaerobic conditions when there is no oxidative phosphorylation to maintain the production of ATP (Adenosine triphosphate) by glycolysis. During fermentation, pyruvate is metabolised to various different compounds. Homolactic fermentation is the production of lactic acid from pyruvate; alcoholic fermentation is the conversion of pyruvate into ethanol and carbon dioxide; and heterolactic fermentation is the production of lactic acid as well as other acids and alcohols. Fermentation does not necessarily have to be carried out in an anaerobic environment. For example, even in the presence of abundant oxygen, yeast cells greatly prefer fermentation to oxidative phosphorylation, as long as sugars are readily available for consumption.

Sugars are the most common substrate of fermentation, and typical examples of fermentation products are ethanol, lactic acid, and hydrogen. However, more exotic compounds can be produced by fermentation, such as butyric acid and acetone. Yeast carries out fermentation in the production of ethanol in beers, wines and other alcoholic drinks, along with the production of large quantities of carbon dioxide.

Fermentation occurs in mammalian muscle during periods of intense exercise where oxygen supply becomes limited, resulting in the creation of lactic acid. Fermentation products contain chemical energy (they are not fully oxidized), but are considered waste products, since they cannot be metabolized further without the use of oxygen (or other more highly-oxidized electron acceptors). A consequence is that the production of adenosine triphosphate (ATP) by fermentation is less efficient than oxidative phosphorylation, whereby pyruvate is fully oxidized to carbon dioxide.

Ethanol Fermentation

Ethanol fermentation (performed by yeast and some types of bacteria) breaks the pyruvate down into ethanol and carbon dioxide. It is important in bread-making, brewing, and wine-making. Usually only one of the products is desired; in bread-making, the alcohol is baked out, and, in alcohol production, the carbon dioxide is released into the atmosphere or used for carbonating the beverage. When the ferment has a high concentration of pectin, minute quantities of methanol can be produced. The chemical equation below summarizes the fermentation of glucose, whose chemical formula is $C_6H_{12}O_6$.

One glucose molecule is converted into two ethanol molecules and two carbon dioxide molecules:

$C_6H_{12}O_6 \rightarrow 2C_2H_5OH + 2CO_2$

C_2H_5OH is the chemical formula for ethanol.

Before fermentation takes place, one glucose molecule is broken down into two pyruvate molecules. This is known as glycolysis.

Lactic Acid Fermentation

Lactic acid fermentation is the simplest type of fermentation. Essentially, it is a redox reaction. In anaerobic conditions, the cell's primary mechanism of ATP production is glycolysis. Glycolysis reduces – transfers electrons to – NAD^+, forming NADH. However, there is only a limited supply of NAD^+ available in a cell. For glycolysis to continue, NADH must be oxidized – have electrons taken away – to regenerate the NAD^+. This is usually done through an electron transport chain in a process called oxidative phosphorylation; however, this mechanism is not available without oxygen.

Instead, the NADH donates its extra electrons to the pyruvate molecules formed during glycolysis. Since the NADH has lost electrons, NAD^+ regenerates and is again available for glycolysis. Lactic acid, for which this process is named, is formed by the reduction of pyruvate. In heterolactic acid fermentation, one molecule of pyruvate is converted to lactate; the other is converted to ethanol and carbon dioxide. In homolactic acid fermentation, both molecules of pyruvate are converted to lactate. Homolactic acid fermentation is unique because it is one of the only respiration processes to not produce a gas as a by-product.

Homolactic fermentation breaks down the pyruvate into lactate. It occurs in the muscles of animals when they need energy faster than the blood can supply oxygen. It also occurs in some kinds of bacteria (such as lactobacilli) and some fungi. It is this type of bacteria that converts lactose into lactic acid in yogurt, giving it its sour taste.

These lactic acid bacteria can be classed as homofermentative, where the end product is mostly lactate, or heterofermentative, where some lactate is further metabolized and results in carbon dioxide, acetate or other metabolic products. The process of lactic acid fermentation using glucose is summarized below. In homolactic fermentation, one molecule of glucose is converted to two molecules of lactic acid:

$C_6H_{12}O_6 \rightarrow 2\ CH_3CHOHCOOH.$

In heterolactic fermentation, the reaction proceeds as follows, with one molecule of glucose being converted to one molecule of lactic acid, one molecule of ethanol, and one molecule of carbon dioxide:

$$C_6H_{12}O_6 \rightarrow CH_3CHOHCOOH + C_2H_5OH + CO_2$$

Before lactic acid fermentation can occur, the molecule of glucose must be split into two molecules of pyruvate. This process is called glycolysis.

Glycolysis

To extract chemical energy from glucose, the glucose molecule must be split into two molecules of pyruvate. This process also generates two molecules of adenosine triphosphate as an immediate energy yield and two molecules of NADH.

$$C_6H_{12}O_6 + 2\ ADP + 2\ P_i + 2\ NAD^+ \rightarrow$$
$$2\ CH_3COCOO^- + 2\ ATP + 2\ NADH + 2\ H_2O + 2H^+$$

The chemical formula of pyruvate is CH_3COCOO^-. P_i stands for the inorganic phosphate. As shown by the reaction equation, glycolysis causes the reduction of two molecules of NAD^+ to NADH. Two ADP molecules are also converted to two ATP and two water molecules via substrate-level phosphorylation.

Aerobic Respiration

In aerobic respiration, the pyruvate produced by glycolysis is further oxidized completely, generating additional ATP and NADH in the citric acid cycle and by oxidative phosphorylation. However, this can only occur in the presence of oxygen. Oxygen is toxic to organisms which are obligate anaerobes, and are not required by facultative anaerobic organisms. In the absence of oxygen, one of the fermentation pathways occurs in order to regenerate NAD^+; lactic acid fermentation is one of these pathways.

Hydrogen Gas Production in Fermentation

Hydrogen gas is produced in many types of fermentation (mixed acid fermentation, butyric acid fermentation, caproate fermentation, butanol fermentation, glyoxylate fermentation), as a way to regenerate NAD^+ from NADH. Electrons are transferred to ferredoxin, which in turn is oxidized by hydrogenase, producing H_2.

Hydrogen gas is a substrate for methanogens and sulfate reducers, which keep the concentration of hydrogen sufficiently low to allow the production of such an energy-rich compound. In fermentation these pyruvic acid molecules are turned into some "waste" product, and a little bit of energy (only two ATP molecules per molecule of glucose – actually four are produced in glycolysis, but two are used up) is produced. Out of many possible types of fermentation processes, two of the most common types are lactic acid fermentation and

Lactic acid fermentation is done by some fungi, some bacteria like the *Lactobacillus acidophilus.* in yogurt, and sometimes by our muscles. Normally our muscles do cellular respiration like the rest of our bodies, using O_2 supplied by our lungs and blood. However, under greater exertion when the oxygen supplied by the lungs and blood system can't get there fast enough to keep up with the muscles' needs, our muscles can switch over and do lactic acid fermentation.

In the process of lactic acid fermentation, the 3-carbon pyruvic acid molecules are turned into lactic acid. It is the presence of lactic acid in yogurt that gives it its sour taste, and it is the presence of lactic acid in our muscles "the morning after" that makes them so sore. Once our muscles form lactic acid, they can't do anything else with it, so until it is gradually washed away by the blood stream and carried to the liver (which is able to get rid of it), our over-exerted muscles feel stiff and sore even if they haven't been physically injured.

Alcohol fermentation is done by yeast and some kinds of bacteria. The "waste" products of this process are ethanol and carbon dioxide (CO_2). Humans have long taken advantage of this process in making bread, beer, and wine. In bread making, it is the CO_2 which forms and is trapped between the gluten (a long protein in wheat) molecules that causes the bread to rise, and the ethanol (often abbreviated as EtOH – do you remember how to draw it?) evaporating that gives it its wonderful smell while baking. The effects of the ethanol in beer and wine are something with which many college students are familiar (sometimes too familiar?), and it is the CO_2 produced by the process of fermentation that makes these beverages effervescent.

Cellular Respiration

An analogy can be drawn between the process of cellular respiration in our cells and a car. The mitochondria are the engines of our cells where sugar is burned for fuel and the exhaust is CO_2 and H_2O. Note that in a car that burned fuel perfectly, the only exhaust should theoretically be CO_2 and H_2O also. There are three steps in the process of cellular respiration: glycolysis, the Krebs cycle, and the electron transport chain.

In contrast to fermentation, in the process of cellular respiration, the pyruvic acid molecules are broken down completely to CO_2 and more energy released. Note that three molecules of O_2 must react with each molecule of pyruvic acid to form the three carbon dioxide molecules, and three molecules of water are also formed to "use up" the hydrogens. As mentioned above, in glycolysis, a total of four molecules of ATP are produced, but two are used up in other steps in the process. Additional ATP is produced during the Krebs Cycle and the Electron Transport Chain, resulting in a grand total of 40 ATP molecules produced from the breakdown of one molecule of glucose via cellular respiration. Since two of those are used up during glycolysis, in prokaryotes a *net* total of 38 molecules of ATP are produced by cellular respiration.

Most prokaryotes have very simple cells which lack several types of organelles present in eukaryotes, and therefore the Krebs Cycle and the Electron Transport Chain occur in the cytoplasm and/or using chemicals embedded in the cell membrane. In contrast, eukaryotes have more complex cells with more specialized organelles to perform given functions. In eukaryotes, the Krebs Cycle and Electron Transport Chain occur within the mitochondria, and thus the pyruvic acid resulting from glycolysis must be sent into the mitochondria for these reactions to occur.

However, to move one molecule of pyruvic acid (remember each molecule of glucose turns into two pyruvic acid molecules) from the cytoplasm into a mitochondrion "costs" the cell one molecule of ATP (therefore two ATPs for a whole glucose), thus a *net* total of 36 ATP molecules per molecule of glucose is produced in eukaryotes as compared to only two in fermentation. The overall reaction for cellular respiration is $C_6H_{12}O_6 + 6O_2 \rightarrow 6CO_2 + 6H_2O$ (+ energy for the cell to use for other things).

In glycolysis and the Krebs cycle, there are also a number of electrons released as the glucose molecule is broken down. The cell must deal with these electrons in some way, so they are stored by the cell by forming a compound called NADH by the chemical reaction, $NAD^+ + H^+ + 2e^-$ NADH. This NADH is used to carry the electrons to the electron transport chain, where more energy is harvested from them.

In eukaryotes, the pyruvic acid from glycolysis must be transferred into the mitochondria to be sent through the Krebs cycle, also known as the citric acid cycle, at a "cost" of one ATP per molecule of pyruvic acid. In this cycle, discovered by Hans Krebs, the pyruvic acid molecules are converted to CO_2, and two more ATP molecules are produced per molecule of glucose. First, each 3-carbon pyruvic acid molecule has a CO_2 broken off and the other two carbons are transferred to a molecule called acetyl coenzyme A, while a molecule of NADH is formed from NAD^+ for each pyruvic acid (= 2 for the whole glucose). These acetyl CoA molecules are put into the actual cycle, and after the coenzyme A part is released, eventually each 2-carbon piece is broken apart into two molecules of CO_2. In the process, for each acetyl CoA that goes into the cycle, three molecules of NADH, one molecule of $FADH_2$, and one molecule of ATP are formed (= 6 NADH, 2 $FADH_2$, and 2 ATP per whole glucose).

The electron transport chain is a system of electron carriers embedded into the inner membrane of a mitochondrion. As electrons are passed from one compound to the next in the chain, their energy is harvested and stored by forming ATP. For each molecule of NADH which puts its two electrons in, approximately three molecules of ATP are formed, and for each molecule of $FADH_2$, about two molecules of ATP are formed.

Many of the compounds that make up the electron transport chain belong to a special group of chemicals called cytochromes. The central structure of a cytochrome is a porphyrin ring like chlorophyll but with iron in the centre (chlorophyll has magnesium). A porphyrin with iron in the centre is called a heme group, and these are also found in hemoglobin in our blood. At the last step in the electron transport chain, the "used up" electrons, along with some "spare" hydrogen ions are combined with O_2 (we finally got around to the O_2) to form water as a waste product: $4e^- + 4H^+ + O_2$ $2H_2O$.

Many of the enzymes in the cells of organisms need other helpers to function. These non-protein enzyme helpers are called cofactors and can include substances like iron, zinc, or copper. If a cofactor is an organic molecule, it then is called a coenzyme. Many of the vitamins needed by our bodies are used as coenzymes to help our enzymes to do their jobs. Vitamin B_1 (thiamine) is a coenzyme used in removing CO_2 from various organic compounds. B_2 (riboflavin) is a component of FAD (or $FADH_2$), one of the chemicals used to transport electrons from the Krebs cycle to the electron transport chain.

Vitamin B_3 (niacin) is a component of NAD^+ (or NADH) which is the major transporter of electrons from glycolysis and the Krebs cycle to the electron transport chain. Without enough of these B vitamins, our ability to get the energy out of our food would come to a grinding halt! B_6 (pyridoxine), B_{12} (cobalamin), pantothenic acid, folic acid, and biotin are all other B vitamins which serve as coenzymes at various points in metabolizing our food. Interestingly, B_{12} has cobalt in it, a mineral which we need in only very minute quantities, but whose absence can cause symptoms of deficiency.

Gluconeogenesis

Gluconeogenesis is the biosynthesis of new glucose, (i.e. not glucose from glycogen). The production of glucose from other metabolites is necessary for use as a fuel source by the brain, testes, erythrocytes and kidney medulla since glucose is the sole energy source for these organs. During starvation, however, the brain can derive energy from ketone bodies which are converted to acetyl-CoA. The primary carbon skeletons used for gluconeogenesis are derived from pyruvate, lactate, glycerol, and the amino acids alanine and glutamine.

The relevant reactions of gluconeogenesis are depicted. The enzymes of the 3 bypass steps are indicated in green along with phosphoglycerate kinase. This latter enzyme is included since when functioning in the gluconeogenic direction the reaction consumes energy. Gluconeogenesis from 2 moles of pyruvate to 2 moles of 1,3-bisphosphoglycerate consumes 6 moles of ATP. This makes the process of gluconeogenesis very costly from an energy standpoint considering that glycolysis to 2 moles of pyruvate only yields 2 moles of ATP. Note that several steps are required in going from 2 moles of 1,3-

bisphosphoglycerate to 1 mole of fructose-1,6-bisphosphate. First there is a reversal of the glyceraldehyde-3-phosphate dehydrogenase reaction which requires a supply of NADH. When lactate is the gluconeogenic substrate the NADH is supplied by the lactate dehydrogenase reaction, and it is supplied by the malate dehydrogenase reaction when pyruvate is the substrate. Secondly, 1 mole of glyceraldehyde-3-phosphate must be isomerized to DHAP and then a mole of DHAP can be condensed to a mole of glyceraldehyde-3-phosphate to form 1 mole of fructose-1,6-bisphosphate in a reversal of the aldolase reaction.

Most non-hepatic tissues lack glucose-6-phosphatase and so the glucose-6-phosphate generated in these tissues would be a substrate for glycogen synthesis. In hepatocytes the glucose-6-phosphatase reactions allows the liver to supply the blood with free glucose. Remember that due to the high K_m of liver glucokinase most of the glucose will not be phosphorylated and will flow down its' concentration gradient out of hepatocytes and into the blood. The three reactions of glycolysis that proceed with a large negative free energy change are bypassed during gluconeogenesis by using different enzymes. These three are the pyruvate kinase, phosphofructokinase-1(PFK-1) and hexokinase/ glucokinase catalysed reactions. In the liver or kidney cortex and in some cases skeletal muscle, the glucose-6-phosphate (G6P) produced by gluconeogenesis can be incorporated into glycogen. In this case the third bypass occurs at the glycogen phosphorylase catalysed reaction. Since skeletal muscle lacks glucose-6-phosphatase it cannot deliver free glucose to the blood and undergoes gluconeogenesis exclusively as a mechanism to generate glucose for storage as glycogen.

Pyruvate to Phosphoenolpyruvate (PEP), Bypass 1

Conversion of pyruvate to PEP requires the action of two mitochondrial enzymes. The first is an ATP-requiring reaction catalysed by pyruvate carboxylase, (PC). As the name of the enzyme implies, pyruvate is carboxylated to form oxaloacetate (OAA). The CO_2 in this reaction is in the form of bicarbonate (HCO_3^-). This reaction is an anaplerotic reaction since it can be used to fill-up the TCA cycle.

The second enzyme in the conversion of pyruvate to PEP is PEP carboxykinase (PEPCK). PEPCK requires GTP in the decarboxylation of OAA to yield PEP. Since PC incorporated CO_2 into pyruvate and it is subsequently released in the PEPCK reaction, no net fixation of carbon occurs. Human cells contain almost equal amounts of mitochondrial and cytosolic PEPCK so this second reaction can occur in either cellular compartment. For gluconeogenesis to proceed, the OAA produced by PC needs to be transported to the cytosol. However, no transport mechanism exist for its' direct transfer and OAA will not freely diffuse. Mitochondrial OAA can become cytosolic via three pathways, conversion to PEP (as indicated above through the action of the mitochondrial PEPCK), transamination to aspartate or reduction to malate, all of which are transported to the cytosol.

If OAA is converted to PEP by mitochondrial PEPCK, it is transported to the cytosol where it is a direct substrate for gluconeogenesis and nothing further is required. Transamination of OAA to aspartate allows the aspartate to be transported to the cytosol where the reverse transamination occurs yielding cytosolic OAA. This transamination reaction requires continuous transport of glutamate into, and α-ketoglutarate out of, the mitochondrion. Therefore, this process is limited by the availability of these other substrates. Either of these latter two reactions will predominate when the substrate for gluconeogenesis is lactate. Whether mitochondrial decarboxylation or transamination occurs is a function of the availability of PEPCK or transamination intermediates.

Mitochondrial OAA can also be reduced to malate in a reversal of the TCA cycle reaction catalysed by malate dehydrogenase (MDH). The reduction of OAA to malate requires NADH, which will be accumulating in the mitochondrion as the energy charge increases. The increased energy charge will allow cells to carry out the ATP costly process of gluconeogenesis. The resultant malate is transported to the cytosol where it is oxidized to OAA by cytosolic MDH which requires NAD^+ and yields NADH. The NADH produced during the cytosolic oxidation of malate to OAA is utilized during the glyceraldehyde-3-phosphate dehydrogenase reaction of glycolysis. The coupling of these two oxidation-reduction reactions is required to keep gluconeogenesis functional when pyruvate is the principal source of carbon atoms. The conversion of OAA to malate predominates when pyruvate (derived from glycolysis or amino acid catabolism) is the source of carbon atoms for gluconeogenesis. When in the cytoplasm, OAA is converted to PEP by the cytosolic version of PEPCK. Hormonal signals control the level of PEPCK protein as a means to regulate the flux through gluconeogenesis.

The net result of the PC and PEPCK reactions is:

$$\text{Pyruvate} + \text{ATP} + \text{GTP} + H_2O \rightarrow$$

$$\text{PEP} + \text{ADP} + \text{GDP} + P_i + 2H^+$$

Fructose-1,6-bisphosphate to Fructose-6-phosphate, Bypass 2

Fructose-1,6-bisphosphate (F1,6BP) conversion to fructose-6-phosphate (F6P) is the reverse of the rate limiting step of glycolysis. The reaction, a simple hydrolysis, is catalysed by fructose-1,6-bisphosphatase (F1,6BPase). Like the regulation of glycolysis occurring at the PFK-1 reaction, the F1,6BPase reaction is a major point of control of gluconeogenesis.

Glucose-6-phosphate (G6P) to Glucose (or Glycogen), Bypass 3

G6P is converted to glucose through the action of glucose-6-phosphatase (G6Pase). This reaction is also a simple hydrolysis reaction like that of F1,6BPase. Since the brain and skeletal muscle, as well as most non-hepatic tissues, lack G6Pase activity, any gluconeogenesis that occurs in these tissues is not utilized for blood glucose supply. In the kidney, muscle and especially the liver, G6P be shunted toward glycogen if blood glucose levels are adequate. The reactions necessary for glycogen synthesis are an alternate bypass series of reactions. Phosphorolysis of glycogen is carried out by glycogen phosphorylase, whereas, glycogen synthesis is catalysed by glycogen synthase. The G6P produced from gluconeogenesis can be converted to glucose-1-phosphate (G1P) by phosphoglucose mutase (PGM). G1P is then converted to UDP-glucose (the substrate for glycogen synthase) by UDP-glucose pyrophosphorylase, a reaction requiring hydrolysis of UTP.

Substrates for Gluconeogenesis

Lactate: Lactate is a predominate source of carbon atoms for glucose synthesis by gluconeogenesis. During anaerobic glycolysis in skeletal muscle, pyruvate is reduced to lactate by lactate dehydrogenase (LDH). This reaction serves two critical functions during anaerobic glycolysis. First, in the direction of lactate formation the LDH reaction requires NADH and yields NAD^+ which is then available for use by the glyceraldehyde-3-phosphate dehydrogenase reaction of glycolysis. These two reaction are, therefore, intimately coupled during anaerobic glycolysis. Secondly, the lactate produced by the LDH reaction is released to the blood stream and transported to the liver where it is converted to glucose. The glucose is then returned to the blood for use by muscle as an energy source and to replenish glycogen stores. This cycle is termed the Cori cycle.

Pyruvate

Pyruvate, generated in muscle and other peripheral tissues, can be transaminated to alanine which is returned to the liver for gluconeogenesis. The transamination reaction requires an α-amino acid as donor of the amino group, generating an α-keto acid in the process. This pathway is termed the glucose-alanine cycle. Although the majority of amino acids are degraded in the liver some are deaminated in muscle.

The glucose-alanine cycle is, therefore, an indirect mechanism for muscle to eliminate nitrogen while replenishing its energy supply. However, the major function of the glucose-alanine cycle is to allow non-hepatic tissues to deliver the amino portion of catabolized amino acids to the liver for excretion as urea. Within the liver the alanine is converted back to pyruvate and used as a gluconeogenic substrate (if that is the hepatic requirement) or oxidized in the TCA cycle. The amino nitrogen is converted to urea in the urea cycle and excreted by the kidneys.

The glucose-alanine cycle is used primarily as a mechanism for skeletal muscle to eliminate nitrogen while replenishing its energy supply. Glucose oxidation produces pyruvate which can undergo transamination to alanine. This reaction is catalysed by alanine transaminase, ALT (ALT used to be referred to a serum glutamate-pyruvate transaminase, SGPT). Additionally, during periods of fasting, skeletal muscle protein is degraded for the energy value of the amino acid carbons and alanine is a major amino acid in protein.

The alanine then enters the blood stream and is transported to the liver. Within the liver alanine is converted back to pyruvate which is then a source of carbon atoms for gluconeogenesis. The newly formed glucose can then enter the blood for delivery back to the muscle. The amino group transported from the muscle to the liver in the form of alanine is converted to urea in the urea cycle and excreted.

Amino Acids

All 20 of the amino acids, excepting leucine and lysine, can be degraded to TCA cycle intermediates as discussed in the metabolism of amino acids. This allows the carbon skeletons of the amino acids to be converted to those in oxaloacetate and subsequently into pyruvate. The pyruvate thus formed can be utilized by the gluconeogenic pathway. When glycogen stores are depleted, in muscle during exertion and liver during fasting, catabolism of muscle proteins to amino acids contributes the major source of carbon for maintenance of blood glucose levels.

Glycerol

Oxidation of fatty acids yields enormous amounts of energy on a molar basis, however, the carbons of the fatty acids cannot be utilized for net synthesis of glucose. The two carbon unit of acetyl-CoA derived from β-oxidation of fatty acids can be incorporated into the TCA cycle, however, during the TCA cycle two carbons are lost as CO_2. Thus, explaining why fatty acids do not undergo net conversion to carbohydrate.

The glycerol backbone of lipids can be used for gluconeogenesis. This requires phosphorylation to glycerol-3-phosphate by glycerol kinase and dehydrogenation to dihydroxyacetone phosphate (DHAP) by glyceraldehyde-3-phosphate dehydrogenase (G3PDH). The G3PDH reaction is the same as that used in the transport of cytosolic reducing equivalents into the mitochondrion for use in oxidative phosphorylation. This transport pathway is called the glycerol-phosphate shuttle.

The glycerol phosphate shuttle is a secondary mechanism for the transport of electrons from cytosolic NADH to mitochondrial carriers of the oxidative phosphorylation pathway. The primary cytoplasmic NADH electron shuttle is the malate-aspartate shuttle. Two enzymes are involved in this shuttle. One is the cytosolic version of the enzyme glycerol-3-phosphate dehydrogenase (glycerol-3-PDH) which has as one substrate, NADH. The second is the mitochondrial form of the enzyme which has as one of its' substrates, FAD^+.

The net result is that there is a continual conversion of the glycolytic intermediate, DHAP and glycerol-3-phosphate with the concomitant transfer of the electrons from reduced cytosolic NADH to mitochondrial oxidized FAD^+. Since the electrons from mitochondrial $FADH_2$ feed into the oxidative phosphorylation pathway at coenzyme Q (as opposed to NADH-ubiquinone oxidoreductase [complex I]) only 2 moles of ATP will be generated from glycolysis. G3PDH is glyceraldehyde-3-phoshate dehydrogenase.

The glycerol backbone of adipose tissue stored triacylgycerols is ensured of being used as a gluconeogenic substrate since adipose cells lack glycerol kinase. In fact adipocytes require a basal level of glycolysis in order to provide them with DHAP as an intermediate in the synthesis of triacyglycerols.

Propionate

Oxidation of fatty acids with an odd number of carbon atoms and the oxidation of some amino acids generates as the terminal oxidation product, propionyl-CoA. Propionyl-CoA is converted to the TCA intermediate, succinyl-CoA. This conversion is carried out by the ATP-requiring enzyme, propionyl-CoA carboxylase then methylmalonyl-CoA epimerase and finally the vitamin B_{12} requiring enzyme, methylmalonyl-CoA mutase. The utilization of propionate in gluconeogenesis only has quantitative significance in ruminants.

Role of Renal Gluconeogenesis

Although the liver has the critical role of maintaining blood glucose homeostasis and therefore, is the major site of gluconeogenesis, the kidney plays an important role. During periods of severe hypoglycemia that occur under conditions of hepatic failure, the kidney can provide glucose to the blood via renal gluconeogenesis. In the renal cortex, glutamine is the preferred substance for gluconeogenesis.

Glutamine is produced in high amounts by skeletal muscle during periods of fasting as a means to export the waste nitrogen resulting from amino acid catabolism. Through the actions of transaminases, a mole of waste ammonia is transferred to α-ketoglutarate via the glutamate dehydrogenase catalysed reaction yielding glutamate. Glutamate is then a substrate for glutamine synthetase which incorporates another mole of waste ammonia generating glutamine. The glutamine is then transported to the kidneys where the reverse reactions occur liberating the ammonia and producing α-ketoglutarate which can enter the TCA cycle and the carbon atoms diverted to gluconeogenesis via oxaloacetate. This process serves two important functions. The ammonia (NH_3) that is liberated spontaneously ionizes to ammonium ion (NH_4^+) and is excreted in the urine effectively buffering the acids in the urine. In addition, the glucose that is produced via gluconeogenesis can provide the brain with critically needed energy.

Regulation of Gluconeogenesis

Obviously the regulation of gluconeogenesis will be in direct contrast to the regulation of glycolysis. In general, negative effectors of glycolysis are positive effectors of gluconeogenesis. Regulation of the activity of PFK-1 and F1,6BPase is the most significant site for controlling the flux toward glucose oxidation or glucose synthesis. As described in control of glycolysis, this is predominantly controlled by fructose-2,6-bisphosphate, F2,6BP which is a powerful negative allosteric effector of F1,6Bpase activity.

Regulation of glycolysis and gluconeogenesis by fructose 2,6-bisphosphate (F2,6BP). The major sites for regulation of glycolysis and gluconeogenesis are the phosphofructokinase-1 (PFK-1) and fructose-1,6-bisphosphatase (F-1,6-BPase) catalysed reactions. PFK-2 is the kinase activity and F-2,6-BPase is the phosphatase activity of the bi-functional regulatory enzyme, phosphofructokinase-2/fructose-2,6-bisphosphatase.

PKA is cAMP-dependent protein kinase which phosphorylates PFK-2/F-2,6-BPase turning on the phosphatase activity. (+ve) and (-ve) refer to positive and negative activities, respectively.

The level of F2,6BP will decline in hepatocytes in response to glucagon stimulation as well as stimulation by catecholamines. Each of these signals is elicited through activation of cAMP-dependent protein kinase (PKA). One substrate for PKA is PFK-2, the bifunctional enzyme responsible for the synthesis and hydrolysis of F2,6BP. When PFK-2 is phosphorylated by PKA it acts as a phosphatase leading to the dephosphorylation of F2,6BP with a concomitant increase in F1,6Bpase activity and a decrease in PFK-1 activity. Secondarily, F1,6Bpase activity is regulated by the ATP/ADP ratio. When this is high, gluconeogenesis can proceed maximally.

Gluconeogenesis is also controlled at the level of the pyruvate to PEP bypass. The hepatic signals elicited by glucagon or epinephrine lead to phosphorylation and inactivation of pyruvate kinase (PK) which will allow for an increase in the flux through gluconeogenesis. PK is also allosterically inhibited by ATP and alanine. The former signals adequate energy and the latter that sufficient substrates for gluconeogenesis are available. Conversely, a reduction in energy levels as evidenced by increasing concentrations of ADP lead to inhibition of both PC and PEPCK. Allosteric activation of PC occurs through acetyl-CoA. Each of these regulations occurs on a short time scale, whereas long-term regulation can be effected at the level of PEPCK. The amount of this enzyme increases in response to prolonged glucagon stimulation. This situation would occur in a starving individual or someone with an inadequate diet.

Whereas glucagon actions results in increased levels of cAMP and subsequent activation of gluconeogenesis, insulin action exerts the opposite effect. The mechanisms by which insulin turns off gluconeogenesis are complex. Reduction in the level of cAMP is exerted via the insulin-mediated activation of phosphodiesterase which hydrolyses cAMP to AMP. At the level of the regulation of genes involved in gluconeogenesis, cAMP signalling leads to phosphorylation of the transcription factor CREB at Ser133.

When phospho-CREB binds to the cAMP response element (CRE) of a target gene it results in the recruitment of the coactivators CBP and p300 (which are closely related). This complex activates gene expression through their intrinsic histone acetyltransferase activity and through recruitment of other coactivator molecules. The coactivator CBP is a target of insulin-dependent phosphorylation at Ser436. This residue, which is adjacent to the CREB-binding domain (CREB-BD), is phosphorylated by a phosphoinositol-3-kinase (PI3K)"dependent insulin signalling pathway, and is not conserved in the related cofactor p300. The protein coactivator PGC-1α (peroxisome proliferator-activated receptor-γ coactivator-1α) functions as a central regulator of gluconeogenesis by binding to several factors, including the glucocorticoid receptor, hepatic nuclear factor 4α (HNF4α) and a forkhead transcription factor (FOXO1) upstream of several genes encoding gluconeogenic enzymes. One mechanism by which insulin signalling antagonizes gluconeogenesis is through phosphorylation of FOXO1 and its subsequent exclusion from the nucleus. Expression of PGC-1α, however, is also regulated by cAMP through a well-defined CRE present in the promoter region.

2

Techniques in Molecular Biology

Molecular biology is the branch of science which deals with the structure and function of macromolecules. It has various techniques which are also applied in other branches of biology.

Here are several techniques used in molecular biology which are also related to fields like genetics, recombinant DNA technology etc. These techniques include following:

Expression cloning - One of the most widely used techniques of molecular biology to study protein function is expression cloning. In this technique DNA coding for a protein of interest is cloned. This is done by using PCR and/or restriction enzymes. The DNA encoding the required protein is inserted into a plasmid which is known as an expression vector.

This plasmid may include special promoter elements to drive production of the protein of interest, and may also have antibiotic resistance markers to help follow the plasmid.

Polymerase Chain Reaction - is most frequently uses technique for copying DNA into millions of clones within few minutes. It can be used to insert restriction enzymes sites in a DNA fragment or to mutate a particular base sequence and also for determination about the presence of a particular DNA fragment in cDNA libraries.

Gel electrophoresis - is used to separate fragments of DNA, RNA and proteins using different properties like according to size, ionic bond etc. For example agarose gel is used to separate RNA and DNA according to their size and SDS-PAGE is used to separate different proteins.

Blotting techniques - is the technique of probing DNA(Southern blotting), RNA(Northern blotting), or proteins(Western blotting) molecules. Eastern blotting is used to recognize post translational modifications of proteins. Other well known and essential techniques in molecular biology include arrays like DNA microarray, allele specific oligonucleotide(ASO) etc.

Since the late 1950s and early 1960s, molecular biologists have learned to characterize, isolate, and manipulate the molecular components of cells and organisms. These components include DNA, the repository of genetic information; RNA, a close relative of DNA whose functions range from serving as a temporary working copy of DNA to actual structural and enzymatic functions as well as a functional and structural part of the translational apparatus; and proteins, the major structural and enzymatic type of molecule in cells.

Expression Cloning

One of the most basic techniques of molecular biology to study protein function is expression cloning. In this technique, DNA coding for a protein of interest is cloned (using PCR and/or restriction enzymes) into a plasmid (known as an expression vector). This plasmid may have special promoter elements to drive production of the protein of interest, and may also have antibiotic resistance markers to help follow the plasmid.

This plasmid can be inserted into either bacterial or animal cells. Introducing DNA into bacterial cells can be done by transformation (via uptake of naked DNA), conjugation (via cell-cell contact) or by transduction (via viral vector). Introducing DNA into eukaryotic cells, such as animal cells, by physical or chemical means is called transfection.

Several different transfection techniques are available, such as calcium phosphate transfection, electroporation, microinjection and liposome transfection. DNA can also be introduced into eukaryotic cells using viruses or bacteria as carriers, the latter is sometimes called bactofection and in particular uses Agrobacterium tumefaciens. The plasmid may be integrated into the genome, resulting in a stable transfection, or may remain independent of the genome, called transient transfection.

In either case, DNA coding for a protein of interest is now inside a cell, and the protein can now be expressed. A variety of systems, such as inducible promoters and specific cell-signalling factors, are available to help express the protein of interest at high levels. Large quantities of a protein can then be extracted from the bacterial or eukaryotic cell. The protein can be tested for enzymatic activity under a variety of situations, the protein may be crystallized so its tertiary structure can be studied, or, in the pharmaceutical industry, the activity of new drugs against the protein can be studied.

Polymerase Chain Reaction (PCR)

The polymerase chain reaction is an extremely versatile technique for copying DNA. In brief, PCR allows a single DNA sequence to be copied (millions of times), or altered in predetermined ways. For example, PCR can be used to introduce restriction enzyme sites, or to mutate (change) particular bases of DNA, the latter is a method referred to as "Quick change". PCR can also be used to determine whether a particular DNA fragment is found in a cDNA library. PCR has many variations, like reverse transcription PCR (RT-PCR) for amplification of RNA, and, more recently, real-time PCR (QPCR) which allow for quantitative measurement of DNA or RNA molecules.

Gel Electrophoresis

Gel electrophoresis is one of the principal tools of molecular biology. The basic principle is that DNA, RNA, and proteins can all be separated by means of an electric field. In agarose gel electrophoresis, DNA and RNA can be separated on the basis of size by running the DNA through an agarose gel. Proteins can be separated on the basis of size by using an SDS-PAGE gel, or on the basis of size and their electric charge by using what is known as a 2D gel electrophoresis.

Macromolecule Blotting and Probing

The terms *northern*, *western* and *eastern* blotting are derived from what initially was a molecular biology joke that played on the term *Southern blotting*, after the technique described by Edwin Southern for the hybridisation of blotted DNA. Patricia Thomas, developer of the RNA blot which then became known as the *northern blot* actually didn't use the term. Further combinations of these techniques produced such terms as *southwesterns* (protein-DNA hybridizations), *northwesterns* (to detect protein-RNA interactions) and *farwesterns* (protein-protein interactions), all of which are presently found in the literature.

Southern Blotting

Named after its inventor, biologist Edwin Southern, the Southern blot is a method for probing for the presence of a specific DNA sequence within a DNA sample. DNA samples before or after restriction enzyme digestion are separated by gel electrophoresis and then transferred to a membrane by blotting via capillary action. The membrane is then exposed to a labeled DNA probe that has a complement base sequence to the sequence on the DNA of interest. Most original protocols used radioactive labels, however non-radioactive alternatives are now available. Southern blotting is less commonly used in laboratory science due to the capacity of other techniques, such as PCR, to detect specific DNA sequences from DNA samples. These blots are still used for some applications, however, such as measuring transgene copy number in transgenic mice, or in the engineering of gene knockout embryonic stem cell lines.

Northern Blotting

The northern blot is used to study the expression patterns of a specific type of RNA molecule as relative comparison among a set of different samples of RNA. It is essentially a combination of denaturing RNA gel electrophoresis, and a blot. In this process RNA is separated based on size and is then transferred to a membrane that is then probed with a labeled complement of a sequence of interest. The results may be visualized through a variety of ways depending on the label used; however, most result in the revelation of bands representing the sizes of the RNA detected in sample. The intensity of these bands is related to the amount of the target RNA in the samples analyzed. The procedure is commonly used to study when and how much gene expression is occurring by measuring how much of that RNA is present in different samples. It is one of the most basic tools for determining at what time, and under what conditions, certain genes are expressed in living tissues.

Western Blotting

Antibodies to most proteins can be created by injecting small amounts of the protein into an animal such as a mouse, rabbit, sheep, or donkey (polyclonal antibodies)or produced in cell culture (monoclonal antibodies). These antibodies can be used for a variety of analytical and preparative techniques.

In western blotting, proteins are first separated by size, in a thin gel sandwiched between two glass plates in a technique known as SDS-PAGE (sodium dodecyl sulfate polyacrylamide gel electrophoresis). The proteins in the gel are then transferred to a PVDF, nitrocellulose, nylon or other support membrane. This membrane can then be probed with solutions of antibodies. Antibodies that specifically bind to the protein of interest can then be visualized by a variety of techniques, including coloured products, chemiluminescence, or autoradiography. Often, the antibodies are labeled with enzymes. When a chemiluminescent substrate is exposed to the enzyme it allows detection. Using western blotting techniques allows not only detection but also quantitative analysis. Analogous methods to western blotting can be used to directly stain specific proteins in live cells or tissue sections. However, these *immunostaining* methods, such as FISH, are used more often in cell biology research.

Eastern Blotting

Eastern blotting technique is to detect post-translational modification of proteins. Proteins blotted on to the PVDF or nitrocellulose membrane are probed for modifications using specific substrates.

Arrays

A DNA array is a collection of spots attached to a solid support such as a microscope slide where each spot contains one or more single-stranded DNA oligonucleotide fragment. Arrays make it possible to put down a large quantity of very small (100 micrometre diameter) spots on a single slide. Each spot has a DNA fragment molecule that is complementary to a single DNA sequence (similar to Southern blotting). A variation of this technique allows the gene expression of an organism at a particular stage in development to be qualified (expression profiling). In this technique the RNA in a tissue is isolated and converted to labeled cDNA.

This cDNA is then hybridized to the fragments on the array and visualization of the hybridization can be done. Since multiple arrays can be made with the exact same position of fragments they are particularly useful for comparing the gene expression of two different tissues, such as a healthy and cancerous tissue. Also, one can measure what genes are expressed and how that expression changes with time or with other factors. For instance, the common baker's yeast, *Saccharomyces cerevisiae*, contains about 7000 genes; with a microarray, one can measure qualitatively how each gene is expressed, and how that expression changes, for example, with a change in temperature.

There are many different ways to fabricate microarrays; the most common are silicon chips, microscope slides with spots of ~ 100 micrometre diameter, custom arrays, and arrays with larger spots on porous membranes (macroarrays). There can be anywhere from 100 spots to more than 10,000 on a given array. Arrays can also be made with molecules other than DNA. For example, an antibody array can be used to determine what proteins or bacteria are present in a blood sample.

Role of Cellular Biology

Main tool used in cellular biology is microscope. A wide range of microscopes including Light microscope, Electron microscope, Fluorescent microscope are used for this purpose. The main purpose of cellular biology is to study different cellular organelles present in all types(prokaryotic-eukaryotic, plant-animal etc) of cells. These organelles include- nucleus, mitochondria, chloroplast, cytoplasm, endoplasmic reticulam, golgi apparatus, cell wall. cell membrane and many more cellular organelles.

Techniques involved in cellular biology other than microscopy include following:

1. Cell culture - in vitro technique of growing cells in laboratory.
2. Immunostaining - is the technique used to locate proteins present in the cell or parts of tissues.

3. Transfection - involve introduction of new gene into a cell
4. Cell fractionation - used to obtain cellular organelles for their study.

Other techniques used in cellular biology include centrifugation, flow cytometry, immunoprecipitation, gene knockdown etc.

Polymerase Chain Reaction

Polymerase chain reaction is a technique of molecular biology which makes multiple copies of DNA from a single fragment. The process of polymerase chain reaction takes place as follows:

1. First the DNA molecule is denatured by heating under the temperature of 94 to 98 degree centigrade.
2. When the strands are separated, they are bound with DNA primers with the temperature 50 to 65 degree Celsius.
3. After annealing, the elongation of the separated DNA strands take place by using heat stable DNA polymerases like Taq (Thermus aquaticus).

There is another method of PCR named reverse transcription PCR which is used to amplify the RNA molecules, while real time PCR is used to measure the DNA o RNA molecule quantitatively.

Expression Cloning

Cloning is the most useful technique of molecular biology because it is used to study the function of the proteins. This technique works in the way that it clones the DNA coding which makes protein, using the process of polymerase chain reaction or restriction enzymes and made multiple copies of that coding. The cloning converts the DNA samples into plasmid vectors. The plasmid either has promoter elements which make possible the production of protein of interest or they have the antibiotic resistant markers.

The plasmid produced from cloning can be inserted in the bacterial cell or the animal cell depending on the mode of experiment. If it is inserted in the bacterial cell, then three processes are involved which make possible the insertion of plasmid into the bacterial cell.

These processes are transformation, conjugation and transduction. When the insertion of the cloned plasmid in the animal cell is concerned then the process of transfection is involved. Transfection techniques are of various types for example, microinjection, calcium phosphate transfection and liposome transfection. But it is also possible that the plasmid can be inserted in the animal cell by bacteria or virus. If bacteria are involved for insertion then it is called as bactofention and mainly the specie of Agrobacterium tumafeciens is used for this purpose.

Plasmid can stay in the eukaryotic cell in two forms; either it mixes up with the genome of the host cell which results in the stable transfection or it stays independently in the cell by transient transfection. In both the cases, the protein of interest is in the cell and can be expressed. If the cell contains specific cell signaling factors or inducible promoters then protein is expressed efficiently than without these factors.

Gel Electrophoresis

Gel electrophoresis is the basic technique of molecular biology which is used to separate DNA, RNA or protein molecules. The method involved is that, the molecules are placed in the immobilized gel and an electric current is passed through it. This current separates the molecules according to the size of the strands. Similarly proteins are also separated in the gel by using an SDS-PAGE gel or it can also be separated according to their size and the charge which they carry.

Allele Specific Oligonucleotide

Allele specific technique is used to detect the single base mutations in the genes. It does not need any gel electrophoresis or polymerase chain reaction. In this technique, short probes are used usually 20 to 25 nucleotides long and they are also labeled. The labeled probes are exposed to the non-fragment target DNA; as a result hybridization takes place due to the short length probes. If there is only a single change in the nucleotide base, then the hybridization will stop. After this step, the target DNA is washed and those probes are removed which did not hybridize. Either radiation or florescence detects the presence of probe on the DNA fragment.

Arrays

This is the technique of molecular biology which uses DNA arrays in the form of spots which are attached to the solid support. This solid support is usually microscopic slide which contains the single stranded DNA oligonucleotide fragment present on each spot. This technique is very feasible because it contains large amount of very short spots with DNA fragments on each slide. The spot is so small that it contains a single DNA fragment. This fragment is complementary to a single DNA sequence. The process starts with the extraction of RNA molecule from a tissue which is converted in the form of cDNA. The DNA fragments present on the spots in the slide are then hybridized with these cDNAs. They can be visualized after hybridization.

This technique is useful in the gene expression data as many arrays can be made with the same position of the DNA fragments on the spots, so that the gene expression of healthy and diseases genes can be compared. For example common yeast consists of 7000 genes. With the microarray, it is possible to observe the expression of each gene qualitatively. It is also possible to view the changes in the expression of these genes due to temperature. There are also other methods of construction of arrays instead of suing DNA fragments. For example an antibody array can be used to determine the presence f proteins or bacteria in the sample of blood.

Protein Folding and Modifications

A polypeptide chain is synthesized on large cellular structures, the ribosomes, by a complex process in which assembly of amino acids in a particular sequence is dictated by messenger RNA (mRNA). The nascent polypeptide chain undergoes folding and, in many cases, chemical modification to generate the final protein. Any polypeptide chain containing *n* residues could, in principle, fold into 8^n conformations.

This value is based on the fact that only eight bond angles are stereochemically allowed in the polypeptide backbone. In general, however, all molecules of any protein species adopt a single conformation, called the *native state,* which is the most stably folded form of the molecule. Misfolding to non-native conformations is suppressed by two distinct mechanisms. At the molecular level, a protein folds through a pathway that favors only a few intermediate steps. Furthermore, a cellular system prevents misfolded proteins from forming. After a protein has carried out its functions, specific sequences that limit the life span of the protein target it for degradation.

The Information for Protein Folding is Encoded in the Sequence

The realization that the amino acid sequence of a protein determines its folding came from in vitro studies on protein unfolding and refolding. Thermal energy from heat, extremes of pH that alter the charges on amino acid side chains, and chemicals such as urea or guanidine hydrochloride at concentrations of 6–8 M can disrupt the weak noncovalent bonds that stabilize the native conformation of a protein.

The denaturation resulting from such treatment causes a protein to lose both its compact conformation and activity. Most denatured proteins precipitate in solution because hydrophobic groups, normally buried inside the molecules, interact with similar regions of other unfolded molecules, causing them to form an insoluble aggregate.

Many proteins that are completely unfolded in 8 M urea and β-mercaptoethanol (which reduces disulfide bonds) can *renature* (refold) into their native state when the denaturing reagents are removed by dialysis. During renaturation, all the disulfide, hydrogen, and hydrophobic bonds that stabilize the native conformation are re-formed. Thus, in this case proteins can be carried through a denaturationrenaturation cycle, which first destroys and then reestablishes their original structure and function. Because renaturation requires no cofactors or other proteins, at least in the test tube, protein folding is a self-assembly process.

Treatment with an 8 M urea solution containing mercaptoethanol ($HSCH_2CH_2OH$) completely denatures most proteins. The urea breaks intramolecular hydrogen and hydrophobic bonds, and the mercaptoethanol reduces each disulfide bridge (–S–S–) to two sulfhydryl (–SH) groups. When these chemicals are removed by dialysis, the –SH groups on the unfolded chain oxidize spontaneously to reform disulfide bridges, and the polypeptide chain simultaneously refolds into its native conformation.

The observation by Christian Anfinsen of such reversible denaturation and renaturation of ribonuclease, an enzyme that degrades RNA, provided a clue that the information for folding a protein lies in its sequence. A general mechanism by which proteins refold in vitro has been elucidated in experiments in which the renaturing conditions are carefully adjusted and the refolding reaction is interrupted at various time intervals.

Such studies have shown that the polypeptide goes through several transient reconfigurations, including a "molten globule" state, before the native tertiary conformation is reached. In the case of ribonuclease, which has several internal disulfide bonds, the folding pathway involves rearrangements of disulfide bond pairs to the native conformation.

Under appropriate refolding conditions, the molecule condenses around a hydrophobic core into a compact, but non-native, intermediate, called a *molten globule.*In this folding intermediate, much of the secondary structure is present. Long-range interactions then form the tertiary structure, folding the molecule into its native three-dimensional conformation.

Protein Synthesis

Cells are capable of synthesizing new proteins, which are essential for the modulation and maintenance of cellular activities. This process involves the formation of new protein molecules from amino acid building blocks based on information encoded in DNA/RNA. Protein synthesis generally consists of two major steps: transcription and translation.

Transcription is the process where genetic information in DNA is used to produce a complementary RNA strand. This RNA strand is then processed to give messenger RNA (mRNA), which is free to migrate through the cell. mRNA molecules bind to protein-RNA complexes called ribosomes located in the cytosol, where they are translated into polypeptide sequences.

The ribosome mediates the formation of a polypeptide sequence based on the mRNA sequence. The mRNA sequence directly relates to the polypeptide sequence by binding to transfer RNA (tRNA) adapter molecules in binding pockets within the ribosome. The new polypeptide then folds into a functional three-dimensional protein molecule.

Movement or Motility

Cells can move during many processes: such as wound healing, the immune response and cancer metastasis. For wound healing to occur, white blood cells and cells that ingest bacteria move to the wound site to kill the microorganisms that cause infection.

At the same time fibroblasts (connective tissue cells) move there to remodel damaged structures. In the case of tumour development, cells from a primary tumour move away and spread to other parts of the body. Cell motility involves many receptors, crosslinking, bundling, binding, adhesion, motor and other proteins. The process is divided into three steps – protrusion of the leading edge of the cell, adhesion of the leading edge and de-adhesion at the cell body and rear, and cytoskeletal contraction to pull the cell forward. Each step is driven by physical forces generated by unique segments of the cytoskeleton.

Evolution

The origin of cells has to do with the origin of life, which began the history of life on Earth.

Origin of the First Cell

There are several theories about the origin of small molecules that could lead to life in an early Earth. One is that they came from meteorites. Another is that they were created at deep-sea vents. A third is that they were synthesized by lightning in a reducing atmosphere; although it is not clear if Earth had such an atmosphere.

There are essentially no experimental data defining what the first self-replicating forms were. RNA is generally assumed to be the earliest self-replicating molecule, as it is capable of both storing genetic information and catalyzing chemical reactions. But some other entity with the potential to self-replicate could have preceded RNA, like clay or peptide nucleic acid.

Cells emerged at least 4.0–4.3 billion years ago. The current belief is that these cells were heterotrophs. An important characteristic of cells is the cell membrane, composed of a bilayer of lipids.

The early cell membranes were probably more simple and permeable than modern ones, with only a single fatty acid chain per lipid. Lipids are known to spontaneously form bilayered vesicles in water, and could have preceded RNA. But the first cell membranes could also have been produced by catalytic RNA, or even have required structural proteins before they could form.

Origin of Eukaryotic Cells

The eukaryotic cell seems to have evolved from a symbiotic community of prokaryotic cells. DNA-bearing organelles like the mitochondria and the chloroplasts are almost certainly what remains of ancient symbiotic oxygen-breathing proteobacteria and cyanobacteria, respectively, where the rest of the cell seems to be derived from an ancestral archaean prokaryote cell – a theory termed the endosymbiotic theory.

There is still considerable debate about whether organelles like the hydrogenosome predated the origin of mitochondria, or viceversa: see the hydrogen hypothesis for the origin of eukaryotic cells.

Sex, as the stereotyped choreography of meiosis and syngamy that persists in nearly all extant eukaryotes, may have played a role in the transition from prokaryotes to eukaryotes. An 'origin of sex as vaccination' theory suggests that the eukaryote genome accreted from prokaryan parasite genomes in numerous rounds of lateral gene transfer.

Sex-as-syngamy (fusion sex) arose when infected hosts began swapping nuclearized genomes containing co-evolved, vertically transmitted symbionts that conveyed protection against horizontal infection by more virulent symbionts.

Folding of Proteins in Vivo is Promoted by Chaperones

Folding of proteins in vitro is an inefficient process, with only a minority of unfolded molecules undergoing complete folding within a few minutes. Clearly, in vivo most protein molecules must rapidly fold into their correct shape; otherwise, cells would waste much energy in the synthesis of nonfunctional proteins and in the degradation of misfolded or unfolded proteins.

More than 95 percent of the proteins present within cells have been shown to be in their native conformation, despite high protein concentrations (H–100 mg/ml), which usually cause proteins to precipitate in vitro.

The explanation for the cell's remarkable efficiency in promoting protein folding probably lies in chaperones, a family of proteins found in all organisms from bacteria to humans. Chaperones are located in every cellular compartment, bind a wide range of proteins, and may be part of a general protein-folding mechanism.

There are two general families of chaperones: *molecular chaperones,* which bind and stabilize unfolded or partially folded proteins, thereby preventing these proteins from being degraded; and *chaperonins,* which directly facilitate their folding. Chaperones have ATPase activity, and their ability to bind and stabilize their target proteins is specific and dependent on ATP hydrolysis. Binding of chaperones to partially folded proteins suggests that the folding process could be regulated at intermediate steps. Molecular chaperones consist of the Hsp70 family of proteins, which includes Hsp70 in the cytosol and mitochondrial matrix, Bip in the endoplasmic reticulum, and DnaK, a bacterial chaperone.

First identified by its rapid appearance after a cell has been stressed by heat shock, Hsp70 is the major chaperone protein in all organisms. When bound to ATP, Hsp70 assumes an open form in which an exposed hydrophobic pocket transiently binds to exposed hydrophobic regions of the unfolded target protein. Hydrolysis of the bound ATP causes Hsp70 to assume a closed form, releasing the target protein. Molecular chaperones are thought to bind all nascent polypeptide chains as they are being synthesized on ribosomes.

In bacteria, 85 percent of the proteins are released from their chaperone and go on to fold normally; an even higher percentage of proteins in eukaryotes follow this pathway.

(a) Many proteins 1 fold into their proper three-dimensional structure with the assistance of Hsp70, a molecular chaperone that transiently binds to a nascent polypeptide as it emerges from a ribosome. Proper folding of some proteins 2 also depends on the chaperonin TCiP, a large barrel-shaped complex of Hsp60 units.

(b) GroEL, the bacterial homolog of TCiP, is a barrel-shaped complex of 14 identical 60,000-MW subunits arranged in two stacked rings. In the absence of ATP or presence of ADP, GroEL exists in a "tight" conformational state that binds partially folded or misfolded proteins. Binding of ATP shifts GroEL to a more open, "relaxed" state, which releases the folded protein.

Proper folding of a small proportion of proteins (e.g., the cytoskeletal proteins actin and tubulin) requires additional assistance, which is provided by chaperonins. Eukaryotic chaperonins, called *TCiP,* are large, barrel-shaped, multimeric complexes composed of eight Hsp60 units.

The bacterial homolog, known as *GroEL,* contains 14 identical subunits. The GroEL folding mechanism, which is better understood than TCiP-mediated folding, serves as a reasonable general model. In bacteria, a partially folded or misfolded polypeptide is inserted into the cavity of GroEL, where it binds to the inner wall and folds into its native conformation.

In an ATP-dependent step, GroEL expands and the protein exits GroEL, a process assisted by a co-chaperonin, GroES, which caps the ends of GroEL. Because the eukaryotic chaperonin TCiP lacks a GroES-type co-chaperonin, the last step must differ in eukaryotes. Moreover, the size of the cavity in TCiP limits this folding pathway to polypeptides smaller than 55 kDa.

Chemical Modifications and Processing Alter the Biological Activity of Proteins

Nearly every protein in a cell is chemically altered after its synthesis on a ribosome. Such modifications may alter the activity, life span, or cellular location of proteins, depending on the nature of the alteration. Protein alterations fall into two categories: chemical modification and processing. *Chemical modification* involves the linkage of a chemical group to the terminal amino or carboxyl groups or to reactive groups in the side chains of internal residues; in some cases, these modifications are reversible. *Processing* involves the removal of peptide segments and generally is irreversible.

Acetylation, the addition of an acetyl group (CH_3CO) to the amino group of the N-terminal residue is the most common form of chemical modification, involving an estimated 80 percent of all proteins:

This modification may play an important role in controlling the life span of proteins within cells, as nonacetylated proteins are rapidly degraded by intracellular proteases. As discussed later, residues at or near the termini of some membrane proteins are chemically modified by addition of long lipidlike groups. Attachment of these hydrophobic "tails," which function to anchor proteins to the lipid bilayer, constitute one way that cells restrict certain proteins to membranes.

The internal residues in proteins can be modified by attachment of a variety of chemical groups to their side chains. The most important modification is *phosphorylation* of serine, threonine, and tyrosine residues. We will encounter numerous examples of proteins whose activity is regulated by reversible phosphorylation and dephosphorylation. The side chains of asparagine, serine, and threonine are sites for *glycosylation,* the attachment of linear and branched carbohydrate chains. Many secreted proteins and membrane proteins contain glycosylated residues. Various less common modifications are found in a limited number of proteins.

Examples of modified internal residues produced by hydroxylation, methylation, and carboxylation. These modifications occur after synthesis of the polypeptide chain.

These modifications occur after synthesis of the polypeptide chain. Unlike chemical modification of residues, which often is reversible, processing of some proteins causes irreversible changes that alter their activity. In the most common form of processing, residues are removed from the C- or N-terminus of a polypeptide by cleavage of the peptide bond in a reaction catalyzed by proteases. Proteolytic cleavage is a common mechanism of activation or inactivation, especially of enzymes involved in blood coagulation or digestion. As discussed later, the activity of certain digestive enzymes is controlled by this mechanism. Proteolysis also generates active peptide hormones, such as EGF mentioned earlier and insulin, from larger precursor polypeptides. An unusual type of processing, termed *protein selfsplicing,* occurs in bacteria and primitive eukaryotes. *Splicing* refers to a process analogous to editing film: an internal segment of polypeptide, an intein, is removed and the ends of the polypeptide are rejoined. Unlike proteolytic processing, protein self-splicing is an autocatalytic process, which proceeds by itself without the involvement of enzymes.

The excised peptide appears to eliminate itself from the protein by a mechanism similar to that used in processing of RNA molecules. In vertebrate cells, processing of some proteins involves self-cleavage, but the subsequent ligation step is absent. One such protein is Hedgehog, which is critical to a number of developmental processes. A segment (red) of a polypeptide, called an *intein,* is removed by cleavage of two peptide bonds (red arrows), leaving two segments (blue and green). One new peptide bond then forms between the two segments (blue arrow), regenerating a continuous polypeptide backbone. The process is autocatalytic and does not depend on enzymes. Consensus splice sites in the polypeptide chain mark the points of breakage and re-formation of the backbone. The excised segment is thought to be exposed at the surface of the folded protein.

Cells Degrade Proteins Via Several Pathways

Cells have both extracellular and intracellular pathways for degrading proteins. The major extracellular pathway is the system of *digestive proteases,* which break down ingested proteins to polypeptides in the intestinal tract. These include *endoproteases* such as trypsin and chymotrypsin, which cleave the protein backbone adjacent to basic and aromatic residues; *exopeptidases,* which sequentially remove residues from the N-terminus (aminopeptidases) or C-terminus (carboxypeptidases) of proteins; and *peptidases,* which split oligopeptides into di- and tripeptides and individual amino acids. These small molecules then are transported across the intestinal lining into the bloodstream.

The life span of intracellular proteins varies from as short as a few minutes for mitotic cyclins, which help regulate passage through mitosis, to as long as the age of an organism for proteins in the lens of the eye. Cells have several intracellular proteolytic pathways for degrading misfolded or denatured proteins, normal proteins whose concentration must be decreased, and foreign proteins taken up by the cell. One major intracellular pathway involves degradation by enzymes within lysosomes, membrane-limited organelles whose interior is acidic. Distinct from the lysosomal pathway are cytosolic mechanisms for degrading proteins. The best-understood pathway, the ubiquitin-mediated pathway, involves two steps: addition of a chain of ubiquitin molecules to an internal lysine side chain of a target protein and proteolysis of the ubiquitinated protein by a proteasome, a large, cylindrical multisubunit complex. The numerous proteasomes present in the cell cytosol proteolytically cleave ubiquitin-tagged proteins in an ATP-dependent process that yields peptides and intact ubiquitin molecules.

A conjugating enzyme catalyzes formation of a peptide bond between ubiquitin (Ub) and the side-chain $–NH_2$ of a lysine residue in a target protein. Additional Ub molecules are added, forming a multiubiquitin chain. This chain is thought to direct the tagged protein to a proteasome, which cleaves the protein into numerous small peptide fragments. (b) Computer-generated image reveals cylindrical structure of a proteasome with a cap at each end of a core. Proteolysis of ubiquitin-tagged proteins occurs along the inner wall of the core.

To be targeted for degradation by the ubiquitinmediated pathway, a protein must contain a structure that is recognized by a ubiquitinating enzyme complex. Different conjugating enzymes recognize different degradation signals in target proteins. For example, the internal sequence Arg-X-X-Leu-Gly-X-Ile-Gly-Asx in mitotic cyclin is recognized by the ubiquitin-conjugating enzyme E1. Internal sequences enriched in proline, glutamic acid, serine, and threonine (PEST sequences) are recognized by other enzymes.

The life span of many cytosolic proteins is correlated with the identity of the N-terminal residue, suggesting that certain residues at the N-terminus favor rapid ubiquitination. For example, short-lived proteins that are degraded within 3 minutes in vivo commonly have Arg, Lys, Phe, Leu, or Trp at their N-terminus. In contrast, a stabilizing amino acid such as Cys, Ala, Ser, Thr, Gly, Val, or Met is present at the N-terminus in long-lived proteins that resist proteolytic attack for more than 30 hours. All newly synthesized proteins have methionine, a stabilizing amino acid, at the N-terminus. Thus subsequent enzymatic alteration that generates one of the destabilizing amino acids at the N-terminus is necessary to target a protein for degradation.

Aberrantly Folded Proteins are Implicated in Slowly Developing Diseases

As noted earlier, each protein species normally folds into a single, energetically favorable conformation that is specified by its amino acid sequence. Recent evidence suggests, however, that a protein may fold into an alternative three- dimensional structure for reasons that have not yet been identified. Such “misfolding” not only leads to a loss of the normal function of a protein but also marks it for proteolytic degradation. The subsequent accumulation of proteolytic fragments contributes to certain degenerative diseases characterized by the presence of insoluble protein plaques in various organs including the liver and brain. Alzheimer’s disease, for example, is marked by formation of plaques and tangles in a deteriorating brain. The filaments composing these structures are derived from proteolytic products of abundant natural proteins such as amyloid precursor protein, a transmembrane protein, and Tau, a microtubule-binding protein. Plaques in other organs are formed from proteolytic fragments of natural proteins such as gelsolin, an actin-binding protein, and serum albumin, a blood protein. The polypeptide fragments liberated by proteolysis polymerize into very stable filaments.

A degeneration of the brain, similar to that seen in Alzheimer's disease, is thought to be caused by *prions,* an infectious protein agent derived by proteolysis and re-folding of a normal brain protein. The amyloid plaque in the brain of an Alzheimer's patient appears as a tangle of filaments. (b) In the atomic force microscope, the filaments are seen to be regular arrangements of a short 47-residue fragment, called β-*amyloid peptide,* produced by proteolysis of amyloid precursor protein.

Summary

- The amino acid sequence of a protein dictates its folding into a specific three-dimensional conformation, the native state.
- Folding of denatured proteins in vitro proceeds through intermediates having secondary and non-native tertiary structure.
- Protein folding in vivo occurs with the assistance of two types of special proteins. Molecular chaperones (Hsp70 proteins) bind to nascent polypeptides emerging from ribosomes and prevent their misfolding. Chaperonins, large complexes of Hsp60-like proteins, shelter some partially folded or misfolded proteins in a barrel-like cavity, providing additional time for proper folding.
- Following their synthesis, all proteins are modified in various ways that alter their structure and function.
- The life span of intracellular proteins is largely determined by their susceptibility to proteolytic degradation by various pathways.
- The presence of certain internal sequences or Nterminal residues targets cytosolic proteins for addition of ubiquitin and subsequent proteolysis within a proteasome.

Genomics and Proteomics

Genomics is a discipline in genetics concerning the study of the genomes of organisms. The field includes intensive efforts to determine the entire DNA sequence of organisms and fine-scale genetic mapping efforts. The field also includes studies of intragenomic phenomena such as heterosis, epistasis, pleiotropy and other interactions between loci and alleles within the genome. In contrast, the investigation of the roles and functions of single genes is a primary focus of molecular biology or genetics and is a common topic of modern medical and biological research. Research of single genes does not fall into the definition of genomics unless the aim of this genetic, pathway, and functional information analysis is to elucidate its effect on, place in, and response to the entire genome's networks.

For the United States Environmental Protection Agency, "the term "genomics" encompasses a broader scope of scientific inquiry associated technologies than when genomics was initially considered. A genome is the sum total of all an individual organism's genes. Thus, genomics is the study of all the genes of a cell, or tissue, at the DNA (genotype), mRNA (transcriptome), or protein (proteome) levels."

History

The first genomes to be sequenced were those of a virus and a mitochondrion, and were done by Fred Sanger. His group established techniques of sequencing, genome mapping, data storage, and bioinformatic analyses in the 1970-1980s. A major branch of genomics is still concerned with sequencing the genomes of various organisms, but the knowledge of full genomes has created the possibility for the field of functional genomics, mainly concerned with patterns of gene expression during various conditions.

The most important tools here are microarrays and bioinformatics. Study of the full set of proteins in a cell type or tissue, and the changes during various conditions, is called proteomics. A related concept is materiomics, which is defined as the study of the material properties of biological materials (e.g. hierarchical protein structures and materials, mineralized biological tissues, etc.) and their effect on the macroscopic function and failure in their biological context, linking processes, structure and properties at multiple scales through a materials science approach. The actual term 'genomics' is thought to have been coined by Dr. Tom Roderick, a geneticist at the Jackson Laboratory (Bar Harbor, ME) over beer at a meeting held in Maryland on the mapping of the human genome in 1986.

In 1972, Walter Fiers and his team at the Laboratory of Molecular Biology of the University of Ghent (Ghent, Belgium) were the first to determine the sequence of a gene: the gene for Bacteriophage MS2 coat protein. In 1976, the team determined the complete nucleotide-sequence of bacteriophage MS2-RNA. The first DNA-based genome to be sequenced in its entirety was that of bacteriophage Φ-X174; (5,368 bp), sequenced by Frederick Sanger in 1977.

The first free-living organism to be sequenced was that of *Haemophilus influenzae* (1.8 Mb) in 1995, and since then genomes are being sequenced at a rapid pace. As of September 2007, the complete sequence was known of about 1879 viruses, 577 bacterial species and roughly 23 eukaryote organisms, of which about half are fungi. Most of the bacteria whose genomes have been completely sequenced are problematic disease-causing agents, such as *Haemophilus influenzae*.

Of the other sequenced species, most were chosen because they were well-studied model organisms or promised to become good models. Yeast (*Saccharomyces cerevisiae*) has long been an important model organism for the eukaryotic cell, while the fruit fly *Drosophila melanogaster* has been a very important tool (notably in early pre-molecular genetics). The worm *Caenorhabditis elegans* is an often used simple model for multicellular organisms.

The zebrafish *Brachydanio rerio* is used for many developmental studies on the molecular level and the flower *Arabidopsis thaliana* is a model organism for flowering plants. The Japanese pufferfish (*Takifugu rubripes*) and the spotted green pufferfish (*Tetraodon nigroviridis*) are interesting because of their small and compact genomes, containing very little non-coding DNA compared to most species. The mammals dog (*Canis familiaris*), brown rat (*Rattus norvegicus*), mouse (*Mus musculus*), and chimpanzee (*Pan troglodytes*) are all important model animals in medical research.

Human Genomics

A rough draft of the human genome was completed by the Human Genome Project in early 2001, creating much fanfare. By 2007 the human sequence was declared "finished" (less than one error in 20,000 bases and all chromosomes assembled). Display of the results of the project required significant bioinformatics resources. The sequence of the human reference assembly can be explored using the UCSC Genome Browser.

Bacteriophage Genomics

Bacteriophages have played and continue to play a key role in bacterial genetics and molecular biology. Historically, they were used to define gene structure and gene regulation. Also the first genome to be sequenced was a bacteriophage. However, bacteriophage research did not lead the genomics revolution, which is clearly dominated by bacterial genomics.

Only very recently has the study of bacteriophage genomes become prominent, thereby enabling researchers to understand the mechanisms underlying phage evolution. Bacteriophage genome sequences can be obtained through direct sequencing of isolated bacteriophages, but can also be derived as part of microbial genomes. Analysis of bacterial genomes has shown that a substantial amount of microbial DNA consists of prophage sequences and prophage-like elements. A detailed database mining of these sequences offers insights into the role of prophages in shaping the bacterial genome.

Cyanobacteria Genomics

At present there are 24 cyanobacteria for which a total genome sequence is available. 15 of these cyanobacteria come from the marine environment. These are six *Prochlorococcus* strains, seven marine *Synechococcus* strains, *Trichodesmium erythraeum* IMS101 and *Crocosphaera watsonii* WH8501. Several studies have demonstrated how these sequences could be used very successfully to infer important ecological and physiological characteristics of marine cyanobacteria.

However, there are many more genome projects currently in progress, amongst those there are further *Prochlorococcus* and marine *Synechococcus* isolates, *Acaryochloris* and *Prochloron*, the N_2-fixing filamentous cyanobacteria *Nodularia spumigena*, *Lyngbya aestuarii* and *Lyngbya majuscula*, as well as bacteriophages infecting marine cyanobaceria.

Thus, the growing body of genome information can also be tapped in a more general way to address global problems by applying a comparative approach. Some new and exciting examples of progress in this field are the identification of genes for regulatory RNAs, insights into the evolutionary origin of photosynthesis, or estimation of the contribution of horizontal gene transfer to the genomes that have been analysed.

Proteomics

Proteomics is the large-scale study of proteins, particularly their structures and functions. Proteins are vital parts of living organisms, as they are the main components of the physiological metabolic pathways of cells. The term "proteomics" was first coined in 1997 to make an analogy with genomics, the study of the genes.

The word "proteome" is a blend of "protein" and "genome", and was coined by Marc Wilkins in 1994 while working on the concept as a PhD student. The proteome is the entire complement of proteins, including the modifications made to a particular set of proteins, produced by an organism or system. This will vary with time and distinct requirements, or stresses, that a cell or organism undergoes.

Complexity of the Problem

After genomics, proteomics is considered the next step in the study of biological systems. It is much more complicated than genomics mostly because while an organism's genome is more or less constant, the proteome differs from cell to cell and from time to time.

This is because distinct genes are expressed in distinct cell types. This means that even the basic set of proteins which are produced in a cell needs to be determined.

In the past this was done by mRNA analysis, but this was found not to correlate with protein content. It is now known that mRNA is not always translated into protein, and the amount of protein produced for a given amount of mRNA depends on the gene it is transcribed from and on the current physiological state of the cell. Proteomics confirms the presence of the protein and provides a direct measure of the quantity present.

Post-translational Modifications

Not only does the translation from mRNA cause differences, many proteins are also subjected to a wide variety of chemical modifications after translation. Many of these post-translational modifications are critical to the protein's function.

Phosphorylation

One such modification is phosphorylation, which happens to many enzymes and structural proteins in the process of cell signaling. The addition of a phosphate to particular amino acids—most commonly serine and threonine mediated by serine/threonine kinases, or more rarely tyrosine mediated by tyrosine kinases—causes a protein to become a target for binding or interacting with a distinct set of other proteins that recognize the phosphorylated domain.

Because protein phosphorylation is one of the most-studied protein modifications, many "proteomic" efforts are geared to determining the set of phosphorylated proteins in a particular cell or tissue-type under particular circumstances. This alerts the scientist to the signaling pathways that may be active in that instance.

Ubiquitination

Ubiquitin is a small protein that can be affixed to certain protein substrates by enzymes called E3 ubiquitin ligases. Determining which proteins are poly-ubiquitinated can be helpful in understanding how protein pathways are regulated. This is therefore an additional legitimate "proteomic" study. Similarly, once it is determined what substrates are ubiquitinated by each ligase, determining the set of ligases expressed in a particular cell type will be helpful.

Additional Modifications

Listing all the protein modifications that might be studied in a "Proteomics" project would require a discussion of most of biochemistry; therefore, a short list will serve here to illustrate the complexity of the problem. In addition to phosphorylation and ubiquitination, proteins can be subjected to (among others) methylation, acetylation, glycosylation, oxidation and nitrosylation. Some proteins undergo ALL of these modifications, often in time-dependent combinations, aptly illustrating the potential complexity one has to deal with when studying protein structure and function.

Distinct Proteins are Made under Distinct Settings

Even if one is studying a particular cell type, that cell may make different sets of proteins at different times, or under different conditions. Furthermore, as mentioned, any one protein can undergo a wide range of post-translational modifications.

Therefore a "proteomics" study can become quite complex very quickly, even if the object of the study is very restricted. In more ambitious settings, such as when a biomarker for a tumour is sought - when the proteomics scientist is obliged to study sera samples from multiple cancer patients - the amount of complexity that must be dealt with is as great as in any modern biological project.

3

The Molecules

Molecules

All matter is made up of tiny particles called *atoms*. Atoms are the basic building blocks of all mater, including life. They cannot be broken down by any known means.

There are only about 100 kinds of atoms but they can be chemically combined to make thousands of different kinds of substances. The substances are called *elements* if they are made up of only one kind of atom.

Hydrogen gas is an example of an element because there are two atoms of Hydrogen joined together. Elements are pure sunstances that cannot be broken down by any oridnay means. Unlike solutions, elements are made up of only one kind of atom with the same atomic number.

For example, the element Nitrogen consists entirely of Nitrogen atoms. There are over 100 different kinds of elements known today. These elements are organized into different families and different periods in the periodic table.

They can be in the form of any state, solid, liquid, or a gas. Compounds are made up of two or more different *elements* put together in definite proportions. For example, water is a compound because it contains two atoms of Hydrogen for every one atom of Oxygen.

A molecule is like a compound in that it is two or more atoms combined, but each atom when combined has to act as a single particle. If a sodium atom passes an electron to a chlorine atom then the chlorine atom has 17 protons and 18 electrons. An atom that has an excess charge is called an ion.

Compounds

Most substances are compounds. A compound is made up of two or more different kinds of atoms combined in definite proportions. Compounds can be separated into the individual atoms that compose their composition, unlike elements that cannot. Compounds can be seperated only by a chemical change that involves heat or electrcity. For example, Water can be separated int Hydrogen and Oxygen by applying a sufficient DC voltage.

Ionic Bonding

Ionic bonding is the bond between two atoms formed as a result of the transfer of electrons from one atom to the other. This results in the formation of positive and negative ions. IONS - Any atom which has an excess charge is called an ion, and the force of attraction between two ions is called an ionic bond. Example: When a sodium atom gives up an electron, it will have 11 protons and 10 electrons. This creates a sodium ion which has an excess positive charge of 1 unit. Relate the concepts of solutions, mixtures, acids and bases to the composition and environment of cells.

A solution is a type of mixture that has the same composition throughtout, and appears to be all one substance. In other words, it's a homogeneous mixture that can be in the form of a solid, liquid or gas. The make up of cells consist of many types of mixtures, including solutions. Many of them are in the form of a liquid. A solution is formed by mixing one or more substances into another substance. In any solution, there is a solute and a solvent. The solvent is generally a liquid, and it is the substance in which the solute is dissolved in. A mixture is a word used to describe a blend of different substances. A mixture is made up of two or more substances, in which they can all be visually seperated and distinguished from one another.

Solutions

A solution is any homogeneous mixture, and may include solids, liquids, as well as gases. Most solutions important to living cells are thought of as liquids. The liquid substance that makes up the bulk of a solution is known as a solvent. Substances that dissolve in the solvent are called solutes (SAHL yoots). A solute can either be a solid, liquid, or gas before they are dissolved in the solvent. The most common solutions have water as the solvent.

Mixtures

A mixture is a combination of substances that are physically mixed without forming new chemical bonds. The substances in a mixture may be present in any proportions. The proportions can change as one substance may be added to, or removed from the mixture, but the different substances in the mixture retain their usual properties.

Acids

An acid is any compound that produces hydrogen ions in a solution. All acids consist of molecules in which contain hydrogen. The molecules are covalently bonded to another atom or group of atoms. When these molecules dissociate in a solution, such as water, the hydrogen breaks loose as an hydrogen ion. The left over molecules form what is called a negative ion. Hydrochloric acid is an example of an acid, which is important to life activities. This is because the level of acid in the environment of the cell affects the actions of enzymes and therefore affects the functioning of the cell.

Base

A base is any compound that produces excess hydroxide ions when it is dissolved in a solution. When in a dry state many bases are ionic compounds. A perfect example of a base is sodium hydroxide. Sodium hydroxide is a solid compound, which is made up of sodium and hydroxide ions. When a base is dissolved in water, it dissociates (separates) into its ions:

$NaOH \longrightarrow Na^{+} + OH^{-}$

An acid is any compound that produces hydrogen ions in solution. All acids consist of molecules that contain hydrogen covalently bonded to another atom or group of atoms.

A base is a compound that produces hydroxide(hy DRAHK syd) ions (OH-) when disolved in water. In the dry slate, many bases are ionic compounds. Sodium hydroxide (NaOH) is one example. This is a solid compound made of sodium and hydroxide ions. When dissolved in water it separates in to it's ions. NaOH then dissociates to form Na+ and OH- ions.

When acid and a base react their hydrogen and hydroxide combine to form water molecules. This reaction removes these two ions from the solution. that is, they are no longer present as separate ions. However, the negative ions of the acid and the positive ions of the base are still present.

Identify the importance of water to life by relating its functions as a transport and dissolving medium. Water is very important to life. Some scientists beleived that life probaly evolved in water. All living cells are made up of 70-95% water and the water covers 75% of the earths surface. Water is present in all three physical states of mater - solid, liquid, and gas.

Water gets its properties from the fact that it is a polar covalent molecule, and therefore forms hydrogen bonds that connects one water molecule to as many as four other water molecules. This results in a property known as cohesion. Water also has adhesive properties to ionic substances as well as other charged substances such as clay. These properties result in many of water's important properties which include its remarkable ability to dissolve a vast variety of covalent and ionic substances, its high heat capacity, capillarity, surface tension, high heat of vaporization, and the fact that water floats when it freezes. All of these properties of water are related to life on Earth.

Water is considered the universal solvent of life. It acts as a solvent which transports molecules and ions from one area of a cell to another. It also acts as a reactant in chemical reactions that occur in the cell. Without water, molecules and ions would be unable to diffuse through the membrane that surrounds the cell, and gases such as carbon dioxide and oxygen would be unable to pass through plasma membranes.

Describe the Characteristics of Organic Compounds

Organic compounds contain carbon atoms that are linked by covalent bonds to form molecular chains, branches, rings, or sheets. Because there are 4 electrons in the outer energy level of the carbon atom, carbon can combine with other carbon atoms, as well as to the many other biologically important elements such as hydrogen, nitrogen, oxygen, sodium, magnesium, phosphorus, sulfur, chlorine, potassium, calcium, and iron. In addition to carbon, the vast majority of organic compounds also contain hydrogen. Oxides, carbonates and cyanides also contain carbon atoms but are not considered to be organic. Such compounds as carbon monoxide, carbon dioxide, calcium carbonate, and cyanide are considered to be inorganic, even though they contain carbon atoms.

Make an analysis of carbohydrates, lipids, proteins, and nucleic acids in terms of: the atoms that compose these compounds, the function these compounds serve in the body, the food sources of each compound, the sub-units that chemically combine to make these compounds, and examples of each group.

The first compound is carbohydrates. Carbohydrates are sugars and starches. Carbohydrates are found in breads, cereals, potatoes, pastas etc. Lipids is the next compound. Lipids are fats, oils and waxes. Lipids are found in mainly fried foods such as pizza, fries, etc. Protein is the next compound. Proteins are compounds that contain nitrogen as well as carbon, hydrogen and oxygen. Proteins are found in beans, fish, and meats. The final compound is nucleic acids. Nucleic acids contain phosphorous and nitrogen along with carbon, hydrogen and oxygen. Nucleic acids are mainly found in meats

Carbohydrates are compunds composed of carbon, hydrogen and oxygen. The name carbohydrates comes from the fact that they are formed from the union of carbon dioxide and water during the process of photosynthesis. The ratio of hydrogen to oxygen in a carbohydrate is always two to one - two hydrogen atoms for every oxygen atom. Carbohydrates are normally foun in the breads and cereals as well as fruits and vegetables.

The simplest carbohydrates are just simple sugars (called monosaccharides). These sugars are very important to the body, because they contain large amounts of energy. This energy can be released in the presence of oxygen by breaking down the sugars into carbon dioxide and water.

Lipids include subtances commonly known as fats, waxes and oils. Like carbohydrates, lipids are composed of carbon, hydrogen and oxygen, except there is less oxygen in lipids, than in carbohydrates. Lipids help make up parts of the cell structure (such as the cell membranes) and are used as a reserve for energy in an organism. Lipids provide twice as much energy per gram as the same amount of carbohydrate. Lipids are found in fatty meats such as bacon, as well as products such as whole milk, cream and butter, and from the oils of many plants such as olives, corn, and peanuts.

Proteins are compuounds containing nitrogen, carbon, hydrogen & oxygen. Some proteins also contain sulfur and phosphorus. The number of proteins in your body is virtually countless. Proteins are found everywhere, in cartilage, bones and muscles, hormones, antibodies, and even enzymes. Proteins are notmally found in meats, poltry, fish and some plants such as beans, soy beans, and legumes.

Nucleic Acids are organic compounds that contain Phosphorus, and Nitrogen, as well as Hydrogen, Oxygen and Carbon. Ther are only two known types of nucleic acids, and they are DNA (Deoxyribonucleic Acid) and RNA (Ribonucleic Acid). These two substances were first found in the Nucleus of a Cell. DNA (which is hereditary) and RNA direct and control the activities and structural devolpment of all the cells of an organism.

Specify the Importance of Nucleic Acids

Nucleic Acids are compounds that contain phosphorous and nitrogen in addition to carbon. There are two types of nucleic acids that are formed by the nucleolus located within the nucleus of cells. DNA (deoxyribonucleic acid) and RNA (ribonucleic acid) carry the genetic code that controls the manufacture or synthesis of proteins. These proteins serve as building blocks of cells, as hormones that regulate cellular function, as antibodies that fight infection, and as enzymes that regulate chemical reactions within the cell.

Nucleic acids are a group of organic substances found in the chromosomes of living cells and viruses. They play a central role in the storage and replication of hereditary information and in the expression of this information through protein synthesis. In most organisms, nucleic acids occur in combination with proteins; the combined substances are called nucleoproteins. Nucleic acid molecules are complex chains of varying length. The two chief types of nucleic acids are DNA (deoxyribonucleic acid), which carries the hereditary information from generation to generation, and RNA (ribonucleic acid), which delivers the instructions coded in this information to the cell's protein manufacturing sites (ribosomes). Outline the function of minerals such as calcium, phosphorus, potassium, and sodium and describe why these substances are essential.

Minerals are chemical elements that organisms need for normal functioning. The most important minerals are iron, calcium, phosphorus, and iodine. Calcium is the most abundant mineral in the human body, and for that matter, the body of any boned animal, as it is the structural element of bones. Weather our body is deficient in calcium depends less on whether we have a sufficient dietary intake, but rather on how well we absorb it. An important factor in calcium absorption is Vitamin D.

The healthy body can produce this vitamin itself for as long as it gets some exposure to sunlight every day. Calcium absorption depends on hormones. Women beyond menopause are under increased risk of developing osteoporosis(bone loss), not because they would lacking in dietary calcium but because they lack the estrogen needed for proper absorption. Trace minerals such as boron also play a part in calcium absorption. The best sources of dietary calcium are milk and milk products, as well as soy and green leafy vegetables. Recommended daily amounts for dietary calcium are about 700 milligram for a healthy adult, and about a gram for adolescents.

Phosphates are essential to the energy-transfer reactions necessary to sustain life processes. Another name for this is adenosine triphospate or ATP which is highly significant to all cells in your body which is involved in every metabolic and photosynthetic reaction. In nature phosphorus occurs only in compounds called phosphates. Plants get phosphorus from soil, and use it in photosynthesis. People and animals take in phosphorus by eating plants and different types of foods.

Distinguish between carbohydrates (sugars and starches), lipids and proteins by using the appropriate tests to identify these nutrients. List each test and how it is conducted.

Tests for lipids: There are two tests that can be used to test for the presence of lipids:

1. Sudan IV : Sudna IV is a special biological stain that is soluble in nonpolar solvents such as lipids and turn from a pink to red colour. Polar compounds will not take up the pink or red colour. For example, if you take a test tube of water and vegetable oil (a lipid) and mix them together and then add a small amount of Sudan IV and shake the mixture, only the oil will turn red or pink not the water.
2. The second test is often called the translucence test for lipid. This is the test that uses unglazed brown paper (same type of brown paper you find in a brown paper bag like the ones from the Co-op used to

rap bottles). If lipid is present, the lipid is absorbed into the paper and causes it to appear translucent which means that light is able to pass through and it looks like a greesy spot.

Tests for Protein: The Biuret Test is often used for protein. As you will learn soon, protein is made of a chain of amino acids that are bonded together by what are known as peptide bonds. The Biuret Test uses Biuret Reagent (a special solution) that is able to react with the peptide bonds of the amino acid chain that makes up the protein. This causes a colour change from blue indicating no protein, to pink (small amount of protein), to violet (larger amount of protein), to purple (largest amount of protein). An example to use is to test a known sample of protein such as egg white or gelatin. Adding the Biuret reagent to either of these food materials will cause a purple colour. Adding Biuret Reagent to water would result in blue colour.

Tests for carbohydrate: Benedicts Reagent is used for single sugars (called monosaccharides). Presence of monosaccharides causes the Benedict's solution to change to various colours when heated. The colour change depends on the concentration of the sugar present. Benedict's Reagent is made from Copper Sulfate and therefore contains Cu^{2+} ions. In the presence of simple sugars (known as reducing sugars) the copper is changed to Cu^{1+} ions. The reaction depends on the concentration of the sugar present. Here is a list of colour changes:

- Blue means no sugar present.
- Light green means 0.5% to 1.0% sugar present.
- Green to Yellow means 1.0% to 1.5% sugar present.
- Orange means 1.5% to 2.0% sugar present.
- Red to red brown means greater than 2.0% sugar present.

Iodine is the test for starch. Mixing iodine and starch causes the formation of a starch-iodine complex that is blue-black in colour. If you place iodine on a sample of food and it remains an amber or light brown colour there is no starch present, but if there is starch present, then the colour will change to a dark blue-black colour.

Explain what vitamins are and describe in general terms the vital body functions of vitamins A, B, C, D, E, and K. Vitamins are organic molecules required in the diet in very small amounts. Vitamins are substances that are normally made by plants and then ingested by consumers. The vitamins are important because they serve as coenzymes which help the enzymes function more effectively.

If the body is unable to supply those necessary enzymes in the proper quantities at the proper time, the vitamins and minerals become inert materials and pass unused through your body. Vitamins function directly as coenzymes, or are converted into coenzymes within the cell. Many important biological reactions need vitamins. Vitamin K and some of the B vitamins are essential for blood clotting. This is because Vitmain K is needed for the symthesis of prothrombin, one of the plasma proteins that takes part in the series of reactions leading to the formation of a blood clott.

When large doses of anitbiotics are given to overcome an infection, they can destroy the intestinal bacteria, and a Vitamin K deficiency may result. This is because Vitamin K is produced as a product of the metabolic activities of these bacteria.

Vitamin A

Vitamin A is a fat-soluble vitamin, which is structurally related to carotene. This is because carotene is converted into vitamin A in the liver. It is composed of compounds called retinoids and carotenoids, and is measured in Retinol Equivalents (RE's) which measure the vitamin A activity of food. There are many vital body functions in which vitamin A is used, some of these are maintaining the protective linings of the lungs, intestine, urinary tract, and other organs, as well as maintaining good vision, and promoting proper bone growth and tooth development.

Too little vitamin A leads to a conditon called night blindness. Night blind people have trouble seeing in dim light. Both black-and-white and colour vision make use of the light-sensitive pigment retinal, which is made from Vitamin A. Others include helping in the formation and maintenance of healthy skin and hair, healthy mucus membranes, fortifying the immune system, reproduction, and hormone synthesis and regulation. Also, Vitamin A is used to treat minor skin disorders, it is important for the human metabolism, and is thought to help prevent the development of cancer. Sources of vitamin A include things such as some diary products, eggs, and liver. Some sources of carotene are orange fruits and vegetables, as well as dark, leafy greens. The absence of Vitamin A in ones diet can cause loss in weight, eye diseases (including night blindness), kidney stones, rashes, depression, frequent infections of the respiratory system, bladder, or digestive systems, plus many other problems.

Vitamin B1 (Thiamine)

Vitamin B1 is a water-soluble vitamin which is important for the production of energy and metabolism of sugar. It is also essential for supporting normal appetite, nervous system functions, muscles and heart, as well as aiding in the digestion of carbohydrates, and promoting growth and good muscle tone. Good sources of vitamin B1 are some dairy products, nuts, meats, whole grains, yeast eggs, and germ of cereals. The lack of vitamin B1 in ones diet may cause the loss of appetite, loss of weight, nausea, anxiety, irritability, insomnia, depression, weakness and fatigue. Other problems that may occur are heart and muscle problems, such as enlarged heart, abnormal heart rhythms, heart failure, muscle wasting, aches and pains in muscles, difficulty walking, loss of reflexes, and paralysis.

Vitamin B2 (Riboflavin)

Vitamin B2 is a water-soluble vitamin which is found to release energy from protein, fat, and carbohydrates, and is the precursor of flavoproteins, flavin-adenine dinucleotide and flavin mononucleotide. This vitamin is part of a coenzyme used in energy metabolism, it alleviates eye fatigue, helps maintain cell respiration, healthy eyes, hair, skin, and nails, aids in the formation of antibodies and red blood cells, and promotes general health.

Good sources of vitamin B2 are some dairy products, liver, the white of an egg, meats, whole grains, cereals, almonds, and green vegetables. Some problems that are caused by the dietary lack of vitamin B2 are magenta tongue, rashes, cracks and sores in or around the mouth, itching, burning, bloodshot eyes, digestive disturbances, trembling, sluggishness, as well as dry, scaly skin.

Vitamin B3 (Niacin)

Vitamin B3 is a water-soluble vitamin, it's precursor is tryptophan, and it is involved in the oxidative release of energy from food. It is necessary for the metabolism of carbohydrates, is part of a coenzyme used in energy metabolism, helps improve circulation, supports the health of the nervous system, digestive system, and the skin, as well as protects the skin.

Other vital body functions of vitamin B3 are preventing pellagra, and reducing high blood pressure, and help lower cholesterol levels. Dietary sources of vitamin B3 include milk, eggs, meat, whole grains and cereals, nuts, and all protein-containing foods.

Insufficient amounts of vitamin B3 in one's diet may result in pellagra, irritability, nervousness, loss of appetite, dizziness, mental confusion, flaky skin rash on sun exposed areas, and delirium. Others include depression, fatigue, headaches, muscular weakness, insomnia, and black, smooth tongue.

Vitamin B5 (Pantothenic Acid)

Vitamin B5 is a water-soluble vitamin which is essential for the metabolism of food. Some vital body functions of vitamin B5 include metabolism of proteins, carbohydrates, and fats, and the production of steroid hormones and other important chemicals, helps fight depression. As well as supporting the normal functioning of the gastrointestinal tract, and it is required for the production of cholesterol, bile, red blood cells, and antibodies. Good dietary sources of vitamin B5 are whole grains, beans, liver, eggs, fish, milk and milk products, yeast, lean beef, some potatoes. Others include broccoli, and other vegetables in the cabbage family. Deficiency of vitamin B5 causes problems such as abdominal distress, burning sensation in the heels, and sleep problems.

Vitamin B6 (Pyridoxine)

Vitamin B6 is a water-soluble vitamin, and is related compounds pyridoxamine, and pyridoxal. This vitamin is necessary for the synthesis and breakdown of amino acids, is used for the metabolism of protein, aids in the formation of red blood cells, helps convert tryptophan to niacin, and for maintaining the central nervous system. It also helps maintain a proper balance of sodium and phosphorous in the body, aids in the removal of excess fluid in premenstrual women, promotes healthy skin, and reduces muscle spasms, leg cramps, nausea, as well as hand numbness and stiffness.

Sources of vitamin B6 include bananas, fish, shellfish, meat, poultry, nuts, fruits, whole grains, vegetables, yeast, wheat germ, maize, and rice husks. The chance of deficiency is increased by smoking, and can cause problems such as anaemia, smooth tongue, insomnia, irritability, muscle twitching, convulsions, rashes, kidney stones, greasy dermatitis. In addition, other resulting problems are arm and leg cramps, loss of hair, muscular weakness, slow learning, mouth disorders, water retention, nervousness, and skin eruptions.

Vitamin B12 (Cyanocobalamin)

Vitamin B12 is a water-soluble vitamin, and is involved in the biosynthesis of methyl groups of choline, and methionine. Some vital body functions of vitamin B12 are being part of a coenzyme used in new cell synthesis, help maintain and protect the nerve cells and system, helps in the formation and regeneration of red blood cells, and helps prevent anaemia.

Others include being used for genetic materials, it is necessary for carbohydrate, fat and protein metabolism, promotes growth in children, increases energy, and is needed for calcium absorption. Significant sources of this vitamin are milk, cheese, eggs, liver and meat.

A shortage of vitamin B12 in one's diet can lead to anaemia, poor appetite, growth failure in children, fatigue, depression, lack of balance, smooth tongue, and skin hypersensitivity. In addition, other problems are brain damage, paralysis, nervous system degeneration, as well as degeneration of the spinal cord.

Define Enzymes and Explain the Make-up and Importance of Enzymes

Enzymes (EN zymz) are protein substances that are necessary for most of the chemical reactions that occur in living cells. Enzymes speed up chemical reactions. Enzymes go into a chemical reaction only temporarily. They are not changed by this reaction so they may be used over and over again for the same chemical step with other molecules. In this way, enzymes are organic catalysts. The substance that an enzyme acts upon is called it's substrate.

Enzymes are almost invisible Protein molecules which are contained in the food we eat and are produced by our bodies. Enzymes are necessary for most of the chemical reations that occur in living cells. All enzymes in an organism are made by the cells of the organism. There are three different types of enzymes which are Metabolic, Digestive, and Food. Enzymes are important because they purify blood, strengthen the immune system, break down fats, lower cholesterol, enhance mental capacity, cleanse the colon, enhance sleep, and improve aging skin.

All living cells make enzymes, but enzymes are not alive. Enzyme molecules function by altering other molecules. Enzymes combine with the altered molecules to form a complex molecule structure in which chemical reactions take place. The enzyme, which remains unchanged, then seprates from the product of the reaction.

The human body has thousands of kinds of enzymes. Each kind does it's own job. Without these people could not breath, see, move, or digest food. Photosynthesis in plants also depends on enzymes. There is a major difference between chemical reactions in nonliving things and the reactions in living things. Consider, for example, the burning of gasoline in an automobile engine. The gasoline vapor is admitted to the engine cylinder. It is ignited by a spark. The vapor burns in a fraction of a second. In fact, the burning is so rapid that it produces a small explosion, which helps to drive the engine.

How Enzymes work is that the ability to act as a catalysts depends on their shape. Somewhere on the surface of each enzyme there is a region called the active site. An Enzyme may also cause 2 molecules to join. In this case, there are two substrates. Each fits into the active site in such a way that they are brought into close contact. This enables bonds to form between the two substrate molecules. There is a theory to explain how enzyme and substrate fit together at an active site. It is called the lock-and-key model. Just as the notched surface of a key can open only one lock, the shape of the active site of an enzyme only fits certian substrates. Thus, each enzyme can catalyze a reaction only of those substrates.

The more modern view of enzyme function is known as the induced fit model which suggests that the enzyme changes shape just slightly to better fit the substrate. Enzymes are proteins which act as biological catalysts. A catalyst enables a chemical reaction to take place at lower temperatures by lowering the activation energy required for the reaction to occur. This is important to living cells since the enzyme enables the metabolic reactions to occur at normal cellular temperatures. Many enzymes must be attached to certain non-protein molecules in order to function. Some molecules are called cofactors, such as copper, iron, and magnesium. Others are called coenzymes. Many coenzymes consist of vitamins, especially B vitamins. If a persons diet lacks the right amount of these vitamins, the enzymes cannot function properly, and serious problems may occur within the body.

Molecules of Life: Biomolecules

A biomolecule is any organic molecule that is produced by a living organism, including large polymeric molecules such as proteins, polysaccharides, and nucleic acids as well as small molecules such as primary metabolites, secondary metabolites, and natural products. A more general name for this class of molecules is a biogenic substance. As organic molecules, biomolecules consist primarily of carbon and hydrogen, nitrogen, and oxygen, and, to a smaller extent, phosphorus and sulfur. Other elements sometimes are incorporated but are much less common.

Types of Biomolecules

A diverse range of biomolecules exist, including:

Small Molecules

In the fields of pharmacology and biochemistry, a small molecule is a low molecular weight organic compound which is by definition not a polymer. The term small molecule, especially within the field of pharmacology, is usually restricted to a molecule that also binds with high affinity to a biopolymer such as protein, nucleic acid, or polysaccharide and in addition alters the activity or function of the biopolymer. The upper molecular weight limit for a small molecule is approximately 800 Daltons which allows for the possibility to rapidly diffuse across cell membranes so that they can reach intracellular sites of action. In addition, this molecular weight cutoff is a necessary but insufficient condition for oral bioavailability.

Small molecules can have a variety of biological functions, serving as cell signaling molecules, as tools in molecular biology, as drugs in medicine, as pesticides in farming, and in many other roles. These compounds can be natural (such as secondary metabolites) or artificial (such as antiviral drugs); they may have a beneficial effect against a disease (such as drugs) or may be detrimental (such as teratogens and carcinogens).

Biopolymers such as nucleic acids, proteins, and polysaccharides (such as starch or cellulose) are not small molecules, although their constituent monomers — ribo- or deoxyribonucleotides, amino acids, and monosaccharides, respectively

— are often considered to be. Very small oligomers are also usually considered small molecules, such as dinucleotides, peptides such as the antioxidant glutathione, and disaccharides such as sucrose.

Drugs

Most drugs are small molecules, although some drugs can be proteins, e.g. insulin. Many proteins are degraded if administered orally and most often cannot cross the cell membranes. Small molecules are more likely to be absorbed, although some of them are only absorbed after oral administration if given as prodrugs.

Many dietary supplements are small molecules (but not herb extracts, such as ginkgo).

Primary and Secondary Metabolites

For organisms to produce small molecules they need one or more specialized enzymes (to create and destroy), which as a result are not that varied in vertebrates (recent and small + slow population size), but very common in soil bacteria (such as streptomyces) and fungi, which in particular secrete antibiotics.

Plants also have several secondary metabolites, which play a role in cell signalling, pigmentation or in defence, several of which have also been used as drugs (medical and recreational):

- Alkaloids
- Glycosides
- Lipids
- Flavonoids
- Nonribosomal peptides, such as actinomycin-D
- Phenazines
- Phenols
- Polyketide
- Terpenes, including steroids
- Tetrapyrroles.

Research Tools

Enzymes and receptors are often activated or inhibited by endogenous protein, but can be also inhibited by endogenous or exogenous small molecule inhibitors or activators with can bind to the active site or on the allosteric site.

An example is the teratogen and carcinogen phorbol 12-myristate 13-acetate which is a plant terpene which activates protein kinase C promoting cancer, making it a very useful investigative tool. There is also interest in creating small molecule artificial transcription factors to regulate gene expression, examples include wrenchnol (a wrench shaped molecule). Binding of ligand can be characterised using a variety of analytical techniques such as surface plasmon resonance, microscale thermophoresis or dual polarisation interferometry to quantify the reaction affinities and kinetic properties and also any induced conformational change.

Nucleosides and Nucleotides

Nucleosides are molecules formed by attaching a nucleobase to a ribose ring. Examples of these include cytidine, uridine, adenosine, guanosine, thymidine and inosine. Nucleosides can be phosphorylated by specific kinases in the cell, producing nucleotides. Both DNA and RNA are polymers, consisting of long, linear molecules. The repeating structural units, or monomers, of the nucleic acids are called nucleotides.

Each nucleotide is made of an acyclic nitrogenous base, a pentose and one to three phosphate groups. They contain carbon, nitrogen, oxygen, hydrogen and phosphorus. They serve as sources of chemical energy (adenosine triphosphate and guanosine triphosphate), participate in cellular signaling (cyclic guanosine monophosphate and cyclic adenosine monophosphate), and are incorporated into important cofactors of enzymatic reactions (coenzyme A, flavin adenine dinucleotide, flavin mononucleotide, and nicotinamide adenine dinucleotide phosphate).

Saccharides

Monosaccharides are the simplest form of carbohydrates with only one simple sugar. They essentially contain an aldehyde or ketone group in their structure. The presence of an aldehyde group in a monosaccharide is indicated by the prefix *aldo*-. Similarly, a ketone group is denoted by the prefix *keto*-. Examples of monosaccharides are the hexoses glucose, fructose, and galactose and pentoses, ribose, and deoxyribose Consumed fructose and glucose have different rates of gastric emptying, are differentially absorbed and have different metabolic fates, providing multiple opportunities for 2 different saccharides to differentially affect food intake.

Disaccharides are formed when two monosaccharides, or two single simple sugars, form a bond with removal of water. They can be hydrolyzed to yield their saccharin building blocks by boiling with dilute acid or reacting them with appropriate enzymes. Examples of disaccharides include sucrose, maltose, and lactose. Polysaccharides are polymerized monosaccharides, complex, carbohydrates. They have multiple simple sugars.

Examples are starch, cellulose, and glycogen. They are generally large and often have a complex branched connectivity. Because of their size, polysaccharides are not water-soluble, but their many hydroxy groups become hydrated individually when exposed to water, and some polysaccharides form thick colloidal dispersions when heated in water. Shorter polysaccharides, with 3 - 10 monomers, are called oligosaccharides. A fluorescent indicator-displacement molecular imprinting sensor was developed for discriminating saccharides. It successfully discriminated three brands of orange juice beverage. The change in fluorescence intensity of the sensing films resulting is directly related to the saccharide concentration.

Lignin

Lignin is a random polymer composed mainly of aromatic rings with short (up to three) aliphatic carbons chains connecting the rings. Lignin is the second most common biopolymer (after cellulose) and is one of the primary structural components of most plants. It contains subunits derived from *p*-coumaryl alcohol, coniferyl alcohol, and sinapyl alcohol and is unusual among biomolecules in that it is racemic i.e. it is not optically active. The lack of optical activity is because the polymerization of lignin occurs via free radical coupling reactions in which there is no preference for either configuration at a chiral center.

Lipids

Lipids are chiefly fatty acid esters, and are the basic building blocks of biological membranes. Another biological role is energy storage (e.g., triglycerides). Most lipids consist of a polar or hydrophilic head (typically glycerol) and one to three nonpolar or hydrophobic fatty acid tails, and therefore they are amphiphilic. Fatty acids consist of unbranched chains of carbon atoms that are connected by single bonds alone (saturated fatty acids) or by both single and double bonds (unsaturated fatty acids). The chains are usually 14-24 carbon groups long, but it is always an even number.

For lipids present in biological membranes, the hydrophilic head is from one of three classes:

- Glycolipids, whose heads contain an oligosaccharide with 1-15 saccharide residues.
- Phospholipids, whose heads contain a positively charged group that is linked to the tail by a negatively charged phosphate group.
- Sterols, whose heads contain a planar steroid ring, for example, cholesterol.

Other lipids include prostaglandins and leukotrienes which are both 20-carbon fatty acyl units synthesized from arachidonic acid. They are also known as fatty acids

Amino Acids

Amino acids contain both amino and carboxylic acid functional groups. (In biochemistry, the term amino acid is used when referring to those amino acids in which the amino and carboxylate functionalities are attached to the same carbon, plus proline which is not actually an amino acid). Amino acids are the building blocks of long polymer chains. With 2-10 amino acids such chains are called peptides, with 10-100 they are often called polypeptides, and longer chains are known as proteins. These protein structures have many structural and enzymatic roles in organisms.

There are twenty amino acids that are encoded by the standard genetic code, but there are more than 500 natural amino acids. When amino acids other than the set of twenty are observed in proteins, this is usually the result of modification after translation (protein synthesis). Only two amino acids other than the standard twenty are known to be incorporated into proteins during translation, in certain organisms:

- Selenocysteine is incorporated into some proteins at a UGA codon, which is normally a stop codon.
- Pyrrolysine is incorporated into some proteins at a UAG codon. For instance, in some methanogens in enzymes that are used to produce methane.

Besides those used in protein synthesis, other biologically important amino acids include carnitine (used in lipid transport within a cell), ornithine, GABA and taurine.

Protein Structure

The particular series of amino acids that form a protein is known as that protein's primary structure. This sequence is determined by the genetic makeup of the individual. Proteins have several, well-classified, elements of local structure formed by intermolecular attraction, this forms the secondary structure of protein. They are broadly divided in two, alpha helix and beta sheet, also called beta pleated sheets. Alpha helices are formed of coiling of protein due to attraction between amine group of one amino acid with carboxylic acid group of other.

The coil contains about 3.6 amino acids per turn and the alkyl group of amino acid lie outside the plane of coil. Beta pleated sheets are formed by strong continuous hydrogen bond over the length of protein chain. Bonding may be parallel or antiparallel in nature.

Structurally, natural silk is formed of beta pleated sheets. Usually, a protein is formed by action of both these structures in variable ratios. Coiling may also be random. The overall 3D structure of a protein is termed its tertiary structure. It is formed as result of various forces like hydrogen bonding, disulfide bridges, hydrophobic interactions, hydrophilic interactions, van der Waals force etc. When two or more different polypeptide chains cluster to form a protein, quaternary structure of protein is formed. Quaternary structure is a unique attribute of polymeric and heteromeric proteins like hemoglobin, which consists of two alpha and two beta peptide chains.

Apoenzymes

An apoenzyme is the inactive storage and generally secretory form of a protein. This is required to protect the secretory cell from the activity of that protein. Apoenzymes becomes active enzyme on addition of a cofactor. Cofactors can be either inorganic (e.g., metal ions and iron-sulfur clusters) or organic compounds, (e.g., flavin and heme). Organic cofactors can be either prosthetic groups, which are tightly bound to an enzyme, or coenzymes, which are released from the enzyme's active site during the reaction.

Isoenzymes

Isoenzymes are enzymes with similar function but different structure. They are products of different genes. They are produced in different organs to perform the same function. LDH are examples of such enzymes. Their varied levels in blood are used to determine any deformity in the organ of secretion.

Vitamins

A vitamin is a compound that is generally not synthesized by a given organism but is nonetheless vital to its survival or health (for example coenzymes). These compounds must be absorbed, or eaten, but typically only in trace quantities. When originally proposed by Casimir Funk, a Polish biochemist, he believed them to all be basic and therefore named them vital amines. The "l" was later dropped to form the word vitamines.

Identification of Genetic Material

The genetic material of a cell or an organism refers to those materials found in the nucleus, mitochondria and cytoplasm, which play a fundamental role in determining the structure and nature of cell substances, and capable of self-propagating and variation.

When it became evident that the chromosomes were the organs of heredity, because (i) they formed the only link between two generations; (ii) they carried linearly arranged genes; (iii) they occurred in every organism in specific number and had specific morphology for a particular species; and (iv)any variation in their number or morphology affected the phenotype of the species.

Then, various attempts were made by early molecular geneticists to identify the physical and chemical nature of genes. But, the genes were found so minute structures that their physical identity remained almost impossible. However, the extensive chemical analysis of chromosomes of different organisms have revealed that chromosomes contain proteins and nucleic acid (DNA and RNA) and it was thought that genes might have either proteins or nucleic acids as their component molecules. Early molecular geneticists have assigned the informational roles of genes to the chromosomal proteins, because, they found nucleic acids too simple to carry genetic informations.

The controversy about the assignment of genetic role either to chromosomal proteins or to chromosomal DNA, existed upto 1950, when finally it was universally accepted that DNA is the genetic substance (I. e., chemical of which genes are composed) of most micro-organisms and higher organisms.Later on, RNA was found to be the genetic material of some viruses. The concept that DNA or RNA is the genetic material of most organisms has been developed and supported by following evidences:

Direct Evidences for DNA as the Genetic Material

The most conclusive evidences in support of DNA as the genetic material comes from following three avenues of approach on micro-organisms: transformation of bacteria mode of infection of bacteriophages and conjugation in bacteria. Bacterial Transformation and Griffith EffectIn 1928, Frederick Griffith encountered a phenomenon now known as genetic transformation. Colonies of virulent strain (pathogenic) of pneumonia causing spherically-shaped bacterium, Diplococcus pneumoniae grown on nutrient agar, have a smooth (s) glistering appearance owing to the presence of a type specific, ploysaccharide (a polymer of glucose and glucuronic acid) capsule.

The avirulant (non-pathogenic) strains of this bacterium, on the other hand, lack this capsule and they produce dull, rough (R) colonies and is unable to cause the pneumonia disease. Smooth (S) and rough (R) characters are directly related to the presence or absence of the capsule and this trait is known to be genetically determined. Both S and R forms occur in several types and are designed as S-1, S-II, S-III, etc., and R-I, R-II, R-III, etc., respectively.

All these subtypes of S and R bacteria differ with each other in the type of antigens, they produce. The kind of antigens produced is likewise genetically determined. Smooth (S) forms sometimes mutate to rough (R) forms, but this change have not been found reversible.

In the course of his work, Griffith injected laboratory mice with live R-II pneumococci; the mice suffered no illness because R-II pneumococci was avirulent.

But when the mice were injected with the mixture of living avirulent R-II and heat killed S-III virulent, the unexpected symptoms of pneumonia appeared and high mortality was resulted in them.

By postmortoming the dead mice, it was found that their heart blood had both R-II and S-III pneumococci. From these results, Griffith concluded that the presence of the heat-killed S-IU bacteria must have caused a transformation of the living R-II bacteria, so as to restore to them the capacity for capsule formation they had earlier lost by gene mutation. This was called "Griffith effect" or more popularly as "bacterial transformation". Identification of the "transforming substance"-Griffith could not understand the cause of bacterial transformation and that is first of all identified by O. T. Avery, C. M. McLeod and M McCarthy (1944). They tested a fraction of heat killed S-III bacteria for the transforming ability.

They could remove proteins, lipids, polysaccharides and ribonucleic acids from the S-III extract by a variety of chemical and enzymatic methods without seriously diminishing its power to transfer R-II mutants into the S-III wild types. They ultimately concluded that because a cell free and highly purified DNA extract of S-III bacteria could bring about transformation of R-I1 bacteria into S-III, therefore, DNA is the genetic material of pneumococci. Later on, such a transforming substance, the DNA, was found in a variety of bacteria (i.e., Hemophilus influenzae. Bacillus subtilus, Shigelia paradysenteriae, etc.) and several other organisms.

Mode of Infection of Bacteriophages and Identification of DNA as their Genetic Material

By using radioactive tracers, A. Hershey and M. Chase (1952) provided further direct proof that DNA is the genetic material in certain bacterial viruses. These investigators were studying the bacteriophages that attack the bacterium, Escherichia coli. They prepared a chemically defined culture medium (for E. coli) containing phosphoric acid and sulphuric acid as the only source of phosphorus and sulphur.

Known quantities of radioactive isotopes of phosphorus (P^{32}) and sulphur (S^{32}) were added to the culture medium. E. coli grown on such a culture medium incorporated the P^{32} and S^{35} into their chemical constituents. When T_2 bacteriophage particles infected the radioactive bacteria, they incorporated the labeled DNA.

Phage capsid proteins did not contain appreciable amounts of phosphorus and only the DNA was labeled with P^{32}. Similarly, the protein envelope around the phage was selectively labeled with S^{35}. DNA contains virtually no sulphur and was not labeled with S^{35} By this method it was possible to differentially label the phage DNA and proteins of phage-capsid.

After phage growth, the phage particles were separated from the host cell by centrifugation. The radioactive phages were next introduced to non-radioactive bacterial cultures where they attacked the bacteria. Subsequently the viruses were separated from the host cell by agitation and the content of

P^{32} and S^{35} in the host (E. coli) and parasite (bacteriophage) was determined. The phosphorus label was found to be associated with the bacterial cells and the sulphur label was in the proteins coats (capsids) left in the medium.

This indicated that the DNA had panetrated the cells but that the protein coat or capsid of the phage was left outside the wall of the bacterium. The labeled DNA was found to reproduced in the host cells. These experiments, thus, clearly demonstrated that genetic informations of these bacteriophages resided in their DNA molecules and not in the protein molecules of their capsids. Later on Hershey-Chase experiment was repeated for other viruses and it was found that DNA is also the genetic material in a number of other viruses.

Bacterial Conjugation

Another conclusive evidence for DNA as the genetic material came from the phenomenon of conjugation of bacteria. Laderberg and Tatum (1946) found that when a F^+ ('male') E. coli cell conjugated with a F^- ('female') E. coli cell an unidirectional transfer of F^+ factor of 'male' cell to F^- or 'female' cell took place; so that the latter was converted into a F^+ or 'male' strain. The F^+ factor was found to be a fragment of DNA molecule which occurred in the cytoplasm of bactarial cell.

Indirect Evidences for DNA as the Genetic Material

The indirect or circumstantial evidences in favour of DNA as the genetic material came from higher organisms which are not as easy to manipulate as bacteria and viruses. The fact that DNA is the genetic material of higher organisms has been supported by following facts:

1. Localization -All cells contain DNA which is localized primarily on chromosomes in the nucleus of the cell. This is consistent with the fact the genes are also located specifically on chromosomes. The Feulgen techniques have shown that DNA entirely remains restricted to the chromosomes and it forms one of the major component of chromosomes.
2. Quantity-Various quantitative measurement of amount of DNA in different cells have shown that there is a correlation between the amount of DNA and the number of chromosome sets (ploidy).

Cells	*Mean DNA Fegulen content (Picograms)*	*Presumed chromosomes set (Ploidy)*
Spermatids	1.68	Haploid(n)
Liver	3.16	Diploid(2n)
Liver	6.30	Tetraploid(4n)
Liver	12.80	Octoploid(8n)

S. No	*Organism*	*Kidney*	*Liver*	*Erythrocytes*	*Sperm*
1.	Chicken	2.4	2.5	2.5	1.3
2.	Bovine	6.4	6.4	-	3.3
3.	Carp	-	3.0	3.3	1.6
4.	Human	5.4	5.6	2.5	2.5

The diploid amount of DNA is constant within the species but varies from one species to another, as shown in.

Stability-Most of the macromolecules of the cell are constantly being broken down and synthesized. If this were to happen to a gene, valuable hereditary information would invariably be lost. Of all the macromolecules in the cell, DNA is the most metabolically stable.

Sensitivity to chemical and physical agents-All agents which cause genes to be damaged (mutation) have also been shown to induce changes in the structure of DNA. For example, one such mutagenic agent, ultraviolet light, is capable of producing breaks in the DNA molecule.

In 1950 Swift found that the amount of DNA per resting nucleus (interphase nucleus) in mitotically active tissues of animals ranged from twice (2n) the value found in the gametes (In) to four times (4n) this value. He concluded that DNA synthesis occurred during the interphase. This parallelism of behaviour in DNA and chromosomes clearly indicates that DNA is the genetic material of higher animals.

Evidences for RNA as the Genetic Material of Some Viruses

The demonstration that RNA is the genetic material in RNA containing viruses came in 1956, when A. Gierer and G. Schramm showed that tobacco plants could be inoculated with purified RNA from the tobacco mosaic virus (TMV) and TMV -like lesions could later be identified on the tobacco leaves. A different approach was taken by H. Fraenkel Conrat and B. Singer in experiments published in 1957. They first separated the RNA from the protein of TMV viruses, much as Hershey and Chase had done with the DNA and the protein of T2 bacteriophage. They then developed techniques for forming 'reconstituted' viruses containing the protein from one mutant strain of TMV and the RNA from another or vice-versa.

Such hybrid viruses were allowed to infect tobacco leaves, and the progeny were examined. In all cases the progeny were the parental RNA type and not the parental protein type. In a much improved technique N. Pace and S. Spiegelman in 1966 purified RNA from two different mutant strains of the RNA phage QB which had quite distinct base compositions. The isolated RNA'S were then incubated separately in the presence of an E. coli cell extract containing an enzyme capable of RNA replication. The new RNA synthesized was in each case identical in base composition to particular phage RNA presented to the in vitro system, thus indicating that the phage RNA-can .serve as a template for its self-replication.

Deoxyribonucleic Acid (DNA)

Deoxyribonucleic acid, or DNA, is a nucleic acid that contains the genetic instructions used in the development and functioning of all known living organisms (with the exception of RNA viruses). The main role of DNA molecules is the long-term storage of information. DNA is often compared to a set of blueprints, like a recipe or a code, since it contains the instructions needed to construct other components of cells, such as proteins and RNA molecules. The DNA segments that carry this genetic information are called genes, but other DNA sequences have structural purposes, or are involved in regulating the use of this genetic information.

DNA consists of two long polymers of simple units called nucleotides, with backbones made of sugars and phosphate groups joined by ester bonds. These two strands run in opposite directions to each other and are therefore anti-parallel. Attached to each sugar is one of four types of molecules called bases. It is the sequence of these four bases along the backbone that encodes information. This information is read using the genetic code, which specifies the sequence of the amino acids within proteins. The code is read by copying stretches of DNA into the related nucleic acid RNA, in a process called transcription.

Within cells, DNA is organized into long structures called chromosomes. These chromosomes are duplicated before cells divide, in a process called DNA replication. Eukaryotic organisms (animals, plants, fungi, and protists) store most of their DNA inside the cell nucleus and some of their DNA in organelles, such as mitochondria or chloroplasts. In contrast, prokaryotes (bacteria and archaea) store their DNA only in the cytoplasm. Within the chromosomes, chromatin proteins such as histones compact and organize DNA. These compact structures guide the interactions between DNA and other proteins, helping control which parts of the DNA are transcribed.

Properties

DNA is a long polymer made from repeating units called nucleotides. As first discovered by James D. Watson and Francis Crick, the structure of DNA of all species comprises two helical chains each coiled round the same axis, and each with a pitch of 34 Ångströms (3.4 nanometres) and a radius of 10 Ångströms (1.0 nanometres). According to another study, when measured in a particular solution, the DNA chain measured 22 to 26 Ångströms wide (2.2 to 2.6 nanometres), and one nucleotide unit measured 3.3 Å (0.33 nm) long. Although each individual repeating unit is very small, DNA polymers can be very large molecules containing millions of nucleotides. For instance, the largest human chromosome, chromosome number 1, is approximately 220 million base pairs long.

In living organisms, DNA does not usually exist as a single molecule, but instead as a pair of molecules that are held tightly together. These two long strands entwine like vines, in the shape of a double helix. The nucleotide repeats contain both the segment of the backbone of the molecule, which holds the chain together, and a base, which interacts with the other DNA strand in the helix. A base linked to a sugar is called a nucleoside and a base linked to a sugar and one or more phosphate groups is called a nucleotide. If multiple nucleotides are linked together, as in DNA, this polymer is called a polynucleotide.

The backbone of the DNA strand is made from alternating phosphate and sugar residues. The sugar in DNA is 2-deoxyribose, which is a pentose (five-carbon) sugar. The sugars are joined together by phosphate groups that form phosphodiester bonds between the third and fifth carbon atoms of adjacent sugar rings. These asymmetric bonds mean a strand of DNA has a direction. In a double helix the direction of the nucleotides in one strand is opposite to their direction in the other strand: the strands are *antiparallel*. The asymmetric ends of DNA strands are called the 52 (*five prime*) and 32 (*three prime*) ends, with the 5' end having a terminal phosphate group and the 3' end a terminal hydroxyl group. One major difference between DNA and RNA is the sugar, with the 2-deoxyribose in DNA being replaced by the alternative pentose sugar ribose in RNA.

The DNA double helix is stabilized primarily by two forces: hydrogen bonds between nucleotides and base-stacking interactions among the aromatic bases. In the aqueous environment of the cell, the conjugated ∂ bonds of nucleotide bases align perpendicular to the axis of the DNA molecule, minimizing their interaction with the solvation shell and therefore, the Gibbs free energy. The four bases found in DNA are adenine (abbreviated A), cytosine (C), guanine (G) and thymine (T). These four bases are attached to the sugar/ phosphate to form the complete nucleotide, as shown for adenosine monophosphate. These bases are classified into two types; adenine and guanine are fused five- and six-membered heterocyclic compounds called purines, while cytosine and thymine are six-membered rings called pyrimidines. A fifth pyrimidine base, called uracil (U), usually takes the place of thymine in RNA and differs from thymine by lacking a methyl group on its ring. Uracil is not usually found in DNA, occurring only as a breakdown product of cytosine. In addition to RNA and DNA, a large number of artificial nucleic acid analogues have also been created to study the proprieties of nucleic acids, or for use in biotechnology.

Grooves

Twin helical strands form the DNA backbone. Another double helix may be found by tracing the spaces, or grooves, between the strands. These voids are adjacent to the base pairs and may provide a binding site. As the strands are not directly opposite each other, the grooves are unequally sized. One groove, the major groove, is 22 Å wide and the other, the minor groove, is 12 Å wide. The narrowness of the minor groove means that the edges of the bases are more accessible in the major groove. As a result, proteins like transcription factors that can bind to specific sequences in double-stranded DNA usually make contacts to the sides of the bases exposed in the major groove. This situation varies in unusual conformations of DNA within the cell, but the major and minor grooves are always named to reflect the differences in size that would be seen if the DNA is twisted back into the ordinary B form.

Base Pair

In molecular biology and genetics, two nucleotides on opposite complementary DNA or RNA strands that are connected via hydrogen bonds are called a base pair (often abbreviated bp). In the canonical Watson-Crick DNA base pairing, adenine (A) forms a base pair with thymine (T) and guanine (G) forms a base pair with cytosine (C). In RNA, thymine is replaced by uracil (U). Alternate hydrogen bonding patterns, such as the wobble base pair and Hoogsteen base pair, also occur—in particular, in RNA—giving rise to complex and functional tertiary structures. It is important to note that pairing is the mechanism by which codons on messenger RNA molecules are recognized by anticodons on transfer RNA during protein translation. Some DNA- or RNA-binding enzymes can recognize specific base pairing patterns that identify particular regulatory regions of genes.

The size of an individual gene or an organism's entire genome is often measured in base pairs because DNA is usually double-stranded. Hence, the number of total base pairs is equal to the number of nucleotides in one of the strands (with the exception of non-coding single-stranded regions of telomeres). The haploid human genome (23 chromosomes) is estimated to be about 3 billion base pairs long and to contain 20,000–25,000 distinct genes. A kilobase (kb) is a unit of measurement in molecular biology equal to 1000 base pairs of DNA or RNA.

Hydrogen Bonding and Stability

Hydrogen bonding is the chemical interaction that underlies the base-pairing rules described above. Appropriate geometrical correspondence of hydrogen bond donors and acceptors allows only the "right" pairs to form stably. DNA with high GC-content is more stable than DNA with low GC-content, but, contrary to popular belief, the hydrogen bonds do not stabilize the DNA significantly, and stabilization is mainly due to stacking interactions. The larger nucleobases, adenine and guanine, are members of a class of double-ringed chemical structures called purines; the smaller nucleobases, cytosine and thymine (and uracil), are members of a class of single-ringed chemical structures called pyrimidines. Purines are complementary only with pyrimidines: pyrimidine-pyrimidine pairings are energetically unfavorable because the molecules are too far apart for hydrogen bonding to be established; purine-purine pairings are energetically unfavorable because the molecules are too close, leading to overlap repulsion.

The only other possible pairings are GT and AC; these pairings are mismatches because the pattern of hydrogen donors and acceptors do not correspond. The GU pairing, with two hydrogen bonds, does occur fairly often in RNA. Paired DNA and RNA molecules are comparatively stable at room temperature but the two nucleotide strands will separate above a melting point that is determined by the length of the molecules, the extent of mispairing (if any), and the GC content. Higher GC content results in higher melting temperatures; it is, therefore, unsurprising that the genomes of extremophile organisms such as *Thermus thermophilus* are particularly GC-rich. On the converse, regions of a genome that need to separate frequently — for example, the promoter regions for often-transcribed genes — are comparatively GC-poor. GC content and melting temperature must also be taken into account when designing primers for PCR reactions.

Base Stacking

Base stacking interactions in DNA and RNA are due to dispersion attraction, short-range exchange repulsion, and electrostatic interactions, which also contribute to stability. Again, GC stacking interactions with adjacent bases tend to be more favorable. (Note, however, that a GC stacking interaction with the next base pair is geometrically different from a CG interaction.) Base stacking effects are especially important in the secondary structure and tertiary structure of RNA; for example, RNA stem-loop structures are stabilized by base stacking in the loop region.

Nucleic Acid Analogues

Nucleic acid analogues are compounds structurally similar (analog) to naturally occurring RNA and DNA, used in medicine and in molecular biology research. Nucleic acids are chains of nucleotides, which are composed of three parts: a phosphate backbone, a pucker-shaped pentose sugar, either ribose or deoxyribose, and one of four nucleobases. An analogue may have any of these altered. Typically the analogue nucleobases confer, among other things, different base pairing and base stacking proprieties. Examples include universal bases, which can pair with all four canon bases, and phosphate-sugar backbone analogues such as PNA, which affect the properties of the chain (PNA can even form a triple helix).

Artificial nucleic acids include peptide nucleic acid (PNA), Morpholino and locked nucleic acid (LNA), as well as glycol nucleic acid (GNA) and threose nucleic acid (TNA). Each of these is distinguished from naturally-occurring DNA or RNA by changes to the backbone of the molecule.

Medicine

Several nucleoside analogues are used as antiviral or anticancer agents. The viral polymerase incorporates these compounds with non-canon bases. These compounds are activated in the cells by being converted into nucleotides, they are administered as nucleosides since charged nucleotides cannot easily cross cell membranes.

Molecular Biology

Nucleic acid analogues are used in molecular biology for several purposes:

- As a tool to detect particular sequences
- As a tool with resistance to RNA hydrolysis
- As a tool for another purpose, such as sequencing
- Naturally occurring, such as in tRNA
- Investigation of the mechanisms used by enzyme, such as an Enzyme inhibitor
- Investigation of possible scenarios of the origin of life
- Investigation of the structural features of nucleic acids
- Investigation of the possible alternatives to the natural system in Synthetic biology.

Backbone Analogues

Hydrolysis Resistant RNA-analogues

To overcome the fact that ribose's 2' hydroxy group that reacts with the phosphate linked 3' hydroxy group (RNA is too unstable to be used or synthesized reliably), a ribose analogue is used. The most common RNA analogues are locked nucleic acid (LNA), morpholino, and peptide nucleic acid (PNA). These oligonucleotides differ as they have a different backbone sugar but still bind according to Watson and Crick pairing with RNA or DNA, but are immune to nuclease activity (They generally cannot be enzymatically synthesized and can only be produced synthetically).

Other Notable Analogues Used as Tools

Dideoxynucleotides are used in sequencing. These nucleoside triphosphates possess a non-canonical sugar, dideoxyribose, which lacks the 3' hydroxyl group normally present in DNA and therefore cannot bond with the next base. The lack of the 3' hydroxyl group terminates the chain reaction as the DNA polymerases mistake it for a regular deoxyribonucleotide. Another chain-terminating analogue that lacks a 3' hydroxyl and mimics adenosine is called cordycepin. Cordycepin is an anticancer drug that targets RNA replication. Another analogue in sequencing is a nucleobase analogue, 7-deaza-GTP and is used to sequence CG rich regions, instead 7-deaza-ATP is called tubercidin, an antibiotic.

Precursors to the RNA World

RNA may be too complex to be the first nucleic acid, so before the RNA world several simpler nucleic acids that differ in the backbone, such as TNA and GNA and PNA, have been offered as candidates for the first nucleic acids.

Base Analogues

Nucleobase Structure and Nomenclature

Natural bases are divided into two classes depending on their structure: pyrimidine (an heterocyclic aromatic six-membered ring with nitrogen atoms in position 1 and 3) and purine (a pyrimidine (numeration inverted) fused with an imidazole ring, a five-membered ring with 2 nitrogen atoms separated by one carbon (meta), 7,9).

Their main proprieties are base pairing, resulting form 2 or 3 hydrogen bonds between ketone (electron withdrawing group,*ei.* more negatively charged) and amino (electron withdrawing group,*ei.* more positively charged) functional groups, and base stacking, caused by the attraction of the delocalized ð electron clouds of the aromatic ring structure.

Fluorophores

Commonly fluorophores (such as rhodamine or fluorescein) are linked to the ring linked to the sugar (in para) via a flexible arm, presumably extruding from the major groove of the helix. Due to low processivity of the nucleotides linked to bulky adducts such as florophores by taq polymerases, the sequence is typically copied using a nucleotide with an arm and later coupled with a reactive fluorophore (indirect labelling):

- amine reactive: Aminoallyl nucleotide contain a primary amine group on a linker that reacts with the amino-reactive dye such as a cyanine or Alexa Fluor dyes, which contain a reactive leaving group, such as a succinimidyl ester (NHS). (base-pairing amino groups are not affected).

- thiol reactive: thiol containing nucleotides reacts with the fluorophore linked to a reactive leaving group, such as a maleimide.
- biotin linked nucleotides rely on the same indirect labelling principle (+ fluorescent streptavidin) and are used in Affymetrix DNAchips.

Fluorophores find a variety of uses in medicine and biochemistry.

Fluorescent Base Analogues

The most commonly used and commercially available fluorescent base analogue, 2-aminopurine (2-AP), has a high-fluorescence quantum yield free in solution (0.68) that is considerably reduced (appr. 100 times but highly dependent on base sequence) when incorporated into nucleic acids. The emission sensitivity of 2-AP to immediate surroundings is shared by other promising and useful fluorescent base analogues like 3-MI, 6-MI, 6-MAP, pyrrolo-dC (also commercially available), modified and improved derivatives of pyrrolo-dC, furan-modified bases and many other ones.

This sensitivity to the microenvironment have been utilized in studies of e.g. structure and dynamics within both DNA and RNA, dynamics and kinetics of DNA-protein interaction and electron transfer within DNA. A newly developed and very interesting group of fluorescent base analogues that has a fluorescence quantum yield that is nearly insensitive to their immediate surroundings is the tricyclic cytosine family. 1,3-Diaza-2-oxophenothiazine, tC, has a fluorescence quantum yield of approximately 0.2 both in single- and in double-strands irrespective of surrounding bases.

Also the oxo-homologue of tC called tC^O (both commercially available), 1,3-diaza-2-oxophenoxazine, has a quantum yield of 0.2 in double-stranded systems. However, it is somewhat sensitive to surrounding bases in single-strands (quantum yields of 0.14–0.41). The high and stable quantum yields of these base analogues make them very bright, and, in combination with their good base analogue properties (leaves DNA structure and stability next to unperturbed), they are especially useful in fluorescence anisotropy and FRET measurements, areas where other fluorescent base analogues are less accurate. Also, in the same family of cytosine analogues, a FRET-acceptor base analogue, tC_{nitro}, has been developed. Together with tC^O as a FRET-donor this constitutes the first nucleic acid base analogue FRET-pair ever developed. The tC-family has, for example, been used in studies related to polymerase DNA-binding and DNA-polymerization mechanisms.

Natural Non-canon Bases

In a cell, there are several noncanon bases present: CpG islands in DNA (are often methylated), all eukaryotic mRNA (capped with a methyl-7-guanosine), and several bases of rRNAs (are methylated). Often, tRNAs are heavily modified postranscriptionally in order to improve their conformation or base pairing, in particular in/near the anticodon: inosine can base pair with C, U, and even with A, whereas thiouridine (with A) is more specific than uracil (with a purine). Other common tRNA base modifications are pseudouridine (which gives its name to the TØC loop), dihydrouridine (which does not stack as it is not aromatic), queuosine, wyosine, and so forth. Nevertheless these are all modifications to normal bases and are not placed by a polymerase.

Base-pairing

Canonical bases may have either a ketone or an amine group on the carbons surrounding the nitrogen atom furthest away from the glycosidic bond, which allows them to base pair (Watson-Crick base pairing) via hydrogen bonds (amine with ketone, purine with pyrimidine). Adenine and 2-aminoadenine have one/two amine group(s), whereas thymine has two ketone groups, and cytosine and guanine are mixed amine and ketone (inverted in respect to each other).

The precise reason why there are only four nucleotides is debated, but there are several unused possibilities. Furthermore, adenine is not the most stable choice for base pairing: in Cyanophage S-2L diaminopurine (DAP) is used instead of adenine (host evasion). Diaminopurine basepairs perfectly with thymine as it is identical to adenine but has an amine group at position 2 forming 3 intramolecular hydrogen bonds, eliminating the major difference between the two types of basepairs (Weak:A-T and Strong:C-G). This improved stability affects protein-binding ineractions that rely on those differences. Other combination include,

- isoguanine and isocytosine, which have their amine and ketone inverted compared to standard guanine and cytosine, (not used probably as tautomers are problematic for base pairing, but isoC and isoG can be amplified correctly with PCR even in the presence of the 4 canon bases)
- diaminopyrimidine and a xanthine, which bind like 2-aminoadenine and thymine but with inverted structures (not used as xanthine is a deamination product).

However, correct DNA structure can form even when the bases are not paired via hydrogen bonding; that is, the bases pair thanks to hydrophobicity, as studies have shown using DNA isosteres (analogues with same number of atoms), such as the thymine analogue 2,4-difluorotoluene (F) or the adenine analogue 4-methylbenzimidazole (Z). An alternative hydrophobic pair could be isoquinoline, and the pyrrolo[2,3-b]pyridine.

Other noteworthy basepairs:

- Several fluorescent bases have also been made, such as the 2-amino-6-(2-thienyl)purine and pyrrole-2-carbaldehyde base pair.

- Metal coordinated bases, such as two 2,6-bis (ethylthiomethyl) pyridine (SPy) with a silver ion or pyridine-2,6-dicarboxamide (Dipam) and a mondentate pyridine (Py) with a copper ion.
- Universal bases may pair indiscriminately with any other base, but, in general, lower the melting temperature of the sequence considerably; examples include 2'-deoxyinosine (hypoxanthine deoxynucleotide) derivatives, nitroazole analogues, and hydrophobic aromatic non-hydrogen-bonding bases (strong stacking effects). These are used as proof of concept and, in general, are not utilised in degenerate primers (which are a mixture of primers).
- The numbers of possible base pairs is doubled when xDNA is considered. xDNA contains expanded bases, in which a benzene ring has been added, which may pair with canon bases, resulting in four possible base-pairs (8 bases:xA-T,xT-A,xC-G,xG-C, 16 bases if the unused arrangements are used). Another form of benzene added bases is yDNA, in which the base is widened by the benzene.

Metal Base Pairs

In metal base-pairing, the Watson-Crick hydrogen bonds are replaced by the interaction between a metal ion with nucleosides acting as ligands. The possible geometries of the metal that would allow for duplex formation with two bidentate nucleosides around a central metal atom are: tetrahedral, dodecahedral, and square planar. Metal-complexing with DNA can occur by the formation of non-canonical base pairs from natural nucleobases with participation by metal ions and also by the exchanging the hydrogen atoms that are part of the Watson-Crick base pairing by metal ions. Introduction of metal ions into a DNA duplex has shown to have potential magnetic, conducting properties, as well as increased stability.

Metal complexing has been shown to occur between natural nucleobases. A well-documented example is the formation of T-Hg-T, which involves two deportonated thymine nucleobases that are brought together by Hg^{2+} and forms a connected metal-base pair.

This motif does not accommodate stacked Hg^{2+} in a duplex due to an intrastrand hairpin formation process that is favored over duplex formation. Two thymines across from each other in a duplex do not form a Watson-Crick base pair in a duplex; this is an example where a Watson-Crick basepair mismatch is stabilized by the formation of the metal-base pair. Another example of a metal complexing to natural nucleobases is the formation of A-Zn-T and G-Zn-C at high pH; Co^{+2} and Ni^{+2} also form these complexes. These are Watson-Crick base pairs where the divalent cation in coordinated to the nucleobases. The exact binding is debated.

A large variety of artificial nucleobases have been developed for use as metal base pairs. These modified nucleobases exhibit tunable electronic properties, sizes, and binding affinities that can be optimized for a specific metal. For, example a nucleoside modified with a pyridine-2,6-dicarboxylate has shown to bind tightly to Cu^{2+}, whereas other divalent ions are only loosely bound. The tridentate character contributes to this selectivity.

The fourth coordination site on the copper is saturated by an oppositely arranged pyridine nucleobase. The asymmetric metal base pairing system is orthogonal to the Watson-Crick base pairs. Another example of an artificial nucleobase is that with hydroxypyridone nucleobases, which are able to bind Cu^{2+} inside the DNA duplex. Five consecutive copper-hydroxypyridone base pairs were incorporated into a double strand, which were flanked by only one natural nucleobase on both ends. EPR data showed that the distance between copper centers was estimated to be 3.7 ± 0.1 Å, while a natural B-type DNA duplex is only slightly larger (3.4 Å). The appeal for stacking metal ions inside a DNA duplex is the hope to obtain nanoscopic self-assembling metal wires, though this has not been realized yet.

Orthogonal System

It has been proposed and studied both theoretically and experimentally the possibility of implementing an orthogonal system inside cells independent of the cellular genetic material in order to make a completely safe system, with the possible increase in encoding potentials Several groups have focused on different aspects:

- novel backbones and base pairs as discussed above
- XNA replication/transcription polymerases starting generally from T7 RNA polymerase
- rybozomes (16S sequences with altered anti Shine-Dalgarno sequence allowing the translation of only orthogonal mRNA with a matching altered Shine-Dalgarno sequence)
- novel tRNA encoding non-natural aminoacids.

Base Analogs and Intercalators

Chemical analogs of nucleotides can take the place of proper nucleotides and establish non-canonical base-pairing, leading to errors (mostly point mutations) in DNA replication and DNA transcription. This is due to their isosteric chemistry. One common mutagenic base analog is 5-bromouracil, which resembles thymine but can base-pair to guanine in its enol form. Other chemicals, known as DNA intercalators, fit into the gap between adjacent bases on a single strand and induce frameshift mutations by "masquerading" as a base, causing the DNA replication machinery to skip or insert additional nucleotides at the intercalated site. Most intercalators are large polyaromatic compounds and are known or suspected carcinogens. Examples include ethidium bromide and acridine.

Examples

The following DNA sequences illustrate pair double-stranded patterns. By convention, the top strand is written from the 5' end to the 3' end; thus, the bottom strand is written 3' to 5'.

A base-paired DNA sequence:

ATCGATTGAGCTCTAGCG

TAGCTAACTCGAGATCGC

The corresponding RNA sequence, in which uracil is substituted for thymine where uracil takes its place in the RNA strand:

AUCGAUUGAGCUCUAGCG

UAGCUAACUCGAGAUCGC

Length Measurements

The following abbreviations are commonly used to describe the length of a D/RNA molecule:

- bp = base pair(s)—one bp corresponds to circa 3.4 Å of length along the strand
- kb (= kbp) = kilo base pairs = 1,000 bp
- Mb = mega base pairs = 1,000,000 bp
- Gb = giga base pairs = 1,000,000,000 bp.

In case of single stranded DNA/RNA units of nucleotides are used, abbreviated nt (or knt, Mnt, Gnt), as they are not paired. For distinction between units of computer storage and bases kbp, Mbp, Gbp, etc. may be used for basepairs. The length of 16S rDNA for bacteria is 1542 base-pairs in length.

The Centimorgan is also often used to imply distance along a chromosome, but the number of base-pairs it corresponds to varies widely. In the Human genome, it is about 1 million base pairs.

Base Pairing

Each type of base on one strand forms a bond with just one type of base on the other strand. This is called complementary base pairing. Here, purines form hydrogen bonds to pyrimidines, with A bonding only to T, and C bonding only to G. This arrangement of two nucleotides binding together across the double helix is called a base pair. As hydrogen bonds are not covalent, they can be broken and rejoined relatively easily. The two strands of DNA in a double helix can therefore be pulled apart like a zipper, either by a mechanical force or high temperature. As a result of this complementarity, all the information in the double-stranded sequence of a DNA helix is duplicated on each strand, which is vital in DNA replication. Indeed, this reversible and specific interaction between complementary base pairs is critical for all the functions of DNA in living organisms.

The two types of base pairs form different numbers of hydrogen bonds, AT forming two hydrogen bonds, and GC forming three hydrogen bonds. DNA with high GC-content is more stable than DNA with low GC-content, due to the added stability of an additional hydrogen bond.

As a result, it is both the percentage of GC base pairs and the overall length of a DNA double helix that determine the strength of the association between the two strands of DNA. Long DNA helices with a high GC content have stronger-interacting strands, while short helices with high AT content have weaker-interacting strands. In biology, parts of the DNA double helix that need to separate easily, such as the TATAAT Pribnow box in some promoters, tend to have a high AT content, making the strands easier to pull apart. In the laboratory, the strength of this interaction can be measured by finding the temperature required to break the hydrogen bonds, their melting temperature (also called T_m value). When all the base pairs in a DNA double helix melt, the strands separate and exist in solution as two entirely independent molecules. These single-stranded DNA molecules (*ssDNA*) have no single common shape, but some conformations are more stable than others.

Sense and Antisense

A DNA sequence is called "sense" if its sequence is the same as that of a messenger RNA copy that is translated into protein. The sequence on the opposite strand is called the "antisense" sequence. Both sense and antisense sequences can exist on different parts of the same strand of DNA (i.e. both strands contain both sense and antisense sequences). In both prokaryotes and eukaryotes, antisense RNA sequences are produced, but the functions of these RNAs are not entirely clear. One proposal is that antisense RNAs are involved in regulating gene expression through RNA-RNA base pairing.

A few DNA sequences in prokaryotes and eukaryotes, and more in plasmids and viruses, blur the distinction between sense and antisense strands by having overlapping genes. In these cases, some DNA sequences do double duty, encoding one protein when read along one strand, and a second protein when read in the opposite direction along the other strand. In bacteria, this overlap may be involved in the regulation of gene transcription, while in viruses, overlapping genes increase the amount of information that can be encoded within the small viral genome.

Supercoiling

DNA can be twisted like a rope in a process called DNA supercoiling. With DNA in its "relaxed" state, a strand usually circles the axis of the double helix once every 10.4 base pairs, but if the DNA is twisted the strands become more tightly or more loosely wound. If the DNA is twisted in the direction of the helix, this is positive supercoiling, and the bases are held more tightly together.

If they are twisted in the opposite direction, this is negative supercoiling, and the bases come apart more easily. In nature, most DNA has slight negative supercoiling that is introduced by enzymes called topoisomerases. These enzymes are also needed to relieve the twisting stresses introduced into DNA strands during processes such as transcription and DNA replication.

Alternate DNA Structures

DNA exists in many possible conformations that include A-DNA, B-DNA, and Z-DNA forms, although, only B-DNA and Z-DNA have been directly observed in functional organisms. The conformation that DNA adopts depends on the hydration level, DNA sequence, the amount and direction of supercoiling, chemical modifications of the bases, the type and concentration of metal ions, as well as the presence of polyamines in solution.

The first published reports of A-DNA X-ray diffraction patterns— and also B-DNA used analyses based on Patterson transforms that provided only a limited amount of structural information for oriented fibers of DNA. An alternate analysis was then proposed by Wilkins *et al.*, in 1953, for the *in vivo* B-DNA X-ray diffraction/scattering patterns of highly hydrated DNA fibers in terms of squares of Bessel functions.

In the same journal, James D. Watson and Francis Crick presented their molecular modeling analysis of the DNA X-ray diffraction patterns to suggest that the structure was a double-helix.

Although the 'B-DNA form' is most common under the conditions found in cells, it is not a well-defined conformation but a family of related DNA conformations that occur at the high hydration levels present in living cells. Their corresponding X-ray diffraction and scattering patterns are characteristic of molecular paracrystals with a significant degree of disorder.

Compared to B-DNA, the A-DNA form is a wider right-handed spiral, with a shallow, wide minor groove and a narrower, deeper major groove. The A form occurs under non-physiological conditions in partially dehydrated samples of DNA, while in the cell it may be produced in hybrid pairings of DNA and RNA strands, as well as in enzyme-DNA complexes. Segments of DNA where the bases have been chemically modified by methylation may undergo a larger change in conformation and adopt the Z form. Here, the strands turn about the helical axis in a left-handed spiral, the opposite of the more common B form. These unusual structures can be recognized by specific Z-DNA binding proteins and may be involved in the regulation of transcription.

Alternate DNA Chemistry

For a number of years exobiologists have proposed the existence of a shadow biosphere, a postulated microbial biosphere of Earth that uses radically different biochemical and molecular processes than currently known life. One of the proposals was the existence of lifeforms that use arsenic instead of phosphorus in DNA.

A December 2010 NASA press conference revealed that the bacterium GFAJ-1, which has evolved in an arsenic-rich environment, is the first terrestrial lifeform found which may have this ability. The bacterium was found in Mono Lake, east of Yosemite National Park. GFAJ-1 is a rod-shaped extremophile bacterium in the family Halomonadaceae that, when starved of phosphorus, may be capable of incorporating the usually poisonous element arsenic in its DNA. This discovery lends weight to the long-standing idea that extraterrestrial life could have a different chemical makeup from life on Earth. The research was carried out by a team led by Felisa Wolfe-Simon, a geomicrobiologist and geobiochemist, a Postdoctoral Fellow of the NASA Astrobiology Institute with Arizona State University.

Quadruplex Structures

At the ends of the linear chromosomes are specialized regions of DNA called telomeres. The main function of these regions is to allow the cell to replicate chromosome ends using the enzyme telomerase, as the enzymes that normally replicate DNA cannot copy the extreme 32 ends of chromosomes. These specialized chromosome caps also help protect the DNA ends, and stop the DNA repair systems in the cell from treating them as damage to be corrected. In human cells, telomeres are usually lengths of single-stranded DNA containing several thousand repeats of a simple TTAGGG sequence.

These guanine-rich sequences may stabilize chromosome ends by forming structures of stacked sets of four-base units, rather than the usual base pairs found in other DNA molecules. Here, four guanine bases form a flat plate and these flat four-base units then stack on top of each other, to form a stable *G-quadruplex* structure.

These structures are stabilized by hydrogen bonding between the edges of the bases and chelation of a metal ion in the centre of each four-base unit. Other structures can also be formed, with the central set of four bases coming from either a single strand folded around the bases, or several different parallel strands, each contributing one base to the central structure.

In addition to these stacked structures, telomeres also form large loop structures called telomere loops, or T-loops. Here, the single-stranded DNA curls around in a long circle stabilized by telomere-binding proteins.

At the very end of the T-loop, the single-stranded telomere DNA is held onto a region of double-stranded DNA by the telomere strand disrupting the double-helical DNA and base pairing to one of the two strands. This triple-stranded structure is called a displacement loop or D-loop.

Branched DNA can form networks containing multiple branches.

Branched DNA

In DNA fraying occurs when non-complementary regions exist at the end of an otherwise complementary double-strand of DNA. However, branched DNA can occur if a third strand of DNA is introduced and contains adjoining regions able to hybridize with the frayed regions of the pre-existing double-strand. Although the simplest example of branched DNA involves only three strands of DNA, complexes involving additional strands and multiple branches are also possible. Branched DNA can be used in nanotechnology to construct geometric shapes.

Vibration

DNA may carry out low-frequency collective motion as observed by the Raman spectroscopy and analysed with a quasi-continuum model.

Chemical Modifications

Structure of cytosine with and without the 5-methyl group. Deamination converts 5-methylcytosine into thymine.

Base Modifications

The expression of genes is influenced by how the DNA is packaged in chromosomes, in a structure called chromatin. Base modifications can be involved in packaging, with regions that have low or no gene expression usually containing high levels of methylation of cytosine bases. For example, cytosine methylation, produces 5-methylcytosine, which is important for X-chromosome inactivation.

The average level of methylation varies between organisms - the worm *Caenorhabditis elegans* lacks cytosine methylation, while vertebrates have higher levels, with up to 1% of their DNA containing 5-methylcytosine. Despite the importance of 5-methylcytosine, it can deaminate to leave a thymine base, so methylated cytosines are particularly prone to mutations. Other base modifications include adenine methylation in bacteria, the presence of 5-hydroxymethylcytosine in the brain, and the glycosylation of uracil to produce the "J-base" in kinetoplastids.

Damage

DNA can be damaged by many sorts of mutagens, which change the DNA sequence. Mutagens include oxidizing agents, alkylating agents and also high-energy electromagnetic radiation such as ultraviolet light and X-rays. The type of DNA damage produced depends on the type of mutagen. For example, UV light can damage DNA by producing thymine dimers, which are cross-links between pyrimidine bases. On the other hand, oxidants such as free radicals or hydrogen peroxide produce multiple forms of damage, including base modifications, particularly of guanosine, and double-strand breaks. A typical human cell contains about 150,000 bases that have suffered oxidative damage. Of these oxidative lesions, the most dangerous are double-strand breaks, as these are difficult to repair and can produce point mutations, insertions and deletions from the DNA sequence, as well as chromosomal translocations.

Many mutagens fit into the space between two adjacent base pairs, this is called *intercalation*. Most intercalators are aromatic and planar molecules; examples include ethidium bromide, daunomycin, and doxorubicin. In order for an intercalator to fit between base pairs, the bases must separate, distorting the DNA strands by unwinding of the double helix.

This inhibits both transcription and DNA replication, causing toxicity and mutations. As a result, DNA intercalators are often carcinogens, and benzo[*a*]pyrene diol epoxide, acridines, aflatoxin and ethidium bromide are well-known examples. Nevertheless, due to their ability to inhibit DNA transcription and replication, other similar toxins are also used in chemotherapy to inhibit rapidly growing cancer cells.

Biological Functions

DNA usually occurs as linear chromosomes in eukaryotes, and circular chromosomes in prokaryotes. The set of chromosomes in a cell makes up its genome; the human genome has approximately 3 billion base pairs of DNA arranged into 46 chromosomes. The information carried by DNA is held in the sequence of pieces of DNA called genes. Transmission of genetic information in genes is achieved via complementary base pairing.

For example, in transcription, when a cell uses the information in a gene, the DNA sequence is copied into a complementary RNA sequence through the attraction between the DNA and the correct RNA nucleotides. Usually, this RNA copy is then used to make a matching protein sequence in a process called translation, which depends on the same interaction between RNA nucleotides. In alternative fashion, a cell may simply copy its genetic information in a process called DNA replication. The details of these functions are covered in other articles; here we focus on the interactions between DNA and other molecules that mediate the function of the genome.

Genes and Genomes

Genomic DNA is tightly and orderly packed in the process called DNA condensation to fit the small available volumes of the cell. In eukaryotes, DNA is located in the cell nucleus, as well as small amounts in mitochondria and chloroplasts. In prokaryotes, the DNA is held within an irregularly shaped body in the cytoplasm called the nucleoid. The genetic information in a genome is held within genes, and the complete set of this information in an organism is called its genotype. A gene is a unit of heredity and is a region of DNA that influences a particular characteristic in an organism. Genes contain an open reading frame that can be transcribed, as well as regulatory sequences such as promoters and enhancers, which control the transcription of the open reading frame.

In many species, only a small fraction of the total sequence of the genome encodes protein. For example, only about 1.5% of the human genome consists of protein-coding exons, with over 50% of human DNA consisting of non-coding repetitive sequences. The reasons for the presence of so much noncoding DNA in eukaryotic genomes and the extraordinary differences in genome size, or *C-value*, among species represent a long-standing puzzle known as the "C-value enigma". However, DNA sequences that do not code protein may still encode functional non-coding RNA molecules, which are involved in the regulation of gene expression.

Some noncoding DNA sequences play structural roles in chromosomes. Telomeres and centromeres typically contain few genes, but are important for the function and stability of chromosomes. An abundant form of noncoding DNA in humans are pseudogenes, which are copies of genes that have been disabled by mutation. These sequences are usually just molecular fossils, although they can occasionally serve as raw genetic material for the creation of new genes through the process of gene duplication and divergence.

Transcription and Translation

A gene is a sequence of DNA that contains genetic information and can influence the phenotype of an organism. Within a gene, the sequence of bases along a DNA strand defines a messenger RNA sequence, which then defines one or more protein sequences. The relationship between the nucleotide sequences of genes and the amino-acid sequences of proteins is determined by the rules of translation, known collectively as the genetic code. The genetic code consists of three-letter 'words' called *codons* formed from a sequence of three nucleotides (e.g. ACT, CAG, TTT).

In transcription, the codons of a gene are copied into messenger RNA by RNA polymerase. This RNA copy is then decoded by a ribosome that reads the RNA sequence by base-pairing the messenger RNA to transfer RNA, which carries amino acids. Since there are 4 bases in 3-letter combinations, there are 64 possible codons (4^3 combinations). These encode the twenty standard amino acids, giving most amino acids more than one possible codon. There are also three 'stop' or 'nonsense' codons signifying the end of the coding region; these are the TAA, TGA and TAG codons.

Replication

Cell division is essential for an organism to grow, but, when a cell divides, it must replicate the DNA in its genome so that the two daughter cells have the same genetic information as their parent. The double-stranded structure of DNA provides a simple mechanism for DNA replication. Here, the two strands are separated and then each strand's complementary DNA sequence is recreated by an enzyme called DNA polymerase. This enzyme makes the complementary strand by finding the correct base through complementary base pairing, and bonding it onto the original strand. As DNA polymerases can only extend a DNA strand in a 52 to 32 direction, different mechanisms are used to copy the antiparallel strands of the double helix. In this way, the base on the old strand dictates which base appears on the new strand, and the cell ends up with a perfect copy of its DNA.

Interactions with Proteins

All the functions of DNA depend on interactions with proteins. These protein interactions can be non-specific, or the protein can bind specifically to a single DNA sequence. Enzymes can also bind to DNA and of these, the polymerases that copy the DNA base sequence in transcription and DNA replication are particularly important.

DNA-binding Proteins

Structural proteins that bind DNA are well-understood examples of non-specific DNA-protein interactions. Within chromosomes, DNA is held in complexes with structural proteins. These proteins organize the DNA into a compact structure called chromatin. In eukaryotes this structure involves DNA binding to a complex of small basic proteins called histones, while in prokaryotes multiple types of proteins are involved. The histones form a disk-shaped complex called a nucleosome, which contains two complete turns of double-stranded DNA wrapped around its surface.

These non-specific interactions are formed through basic residues in the histones making ionic bonds to the acidic sugar-phosphate backbone of the DNA, and are therefore largely independent of the base sequence. Chemical modifications of these basic amino acid residues include methylation, phosphorylation and acetylation. These chemical changes alter the strength of the interaction between the DNA and the histones, making the DNA more or less accessible to transcription factors and changing the rate of transcription. Other non-specific DNA-binding proteins in chromatin include the high-mobility group proteins, which bind to bent or distorted DNA. These proteins are important in bending arrays of nucleosomes and arranging them into the larger structures that make up chromosomes.

A distinct group of DNA-binding proteins are the DNA-binding proteins that specifically bind single-stranded DNA. In humans, replication protein A is the best-understood member of this family and is used in processes where the double helix is separated, including DNA replication, recombination and DNA repair. These binding proteins seem to stabilize single-stranded DNA and protect it from forming stem-loops or being degraded by nucleases.

In contrast, other proteins have evolved to bind to particular DNA sequences. The most intensively studied of these are the various transcription factors, which are proteins that regulate transcription. Each transcription factor binds to one particular set of DNA sequences and activates or inhibits the transcription of genes that have these sequences close to their promoters.

The transcription factors do this in two ways. Firstly, they can bind the RNA polymerase responsible for transcription, either directly or through other mediator proteins; this locates the polymerase at the promoter and allows it to begin transcription. Alternatively, transcription factors can bind enzymes that modify the histones at the promoter; this will change the accessibility of the DNA template to the polymerase.

As these DNA targets can occur throughout an organism's genome, changes in the activity of one type of transcription factor can affect thousands of genes. Consequently, these proteins are often the targets of the signal transduction processes that control responses to environmental changes or cellular differentiation and development. The specificity of these transcription factors' interactions with DNA come from the proteins making multiple contacts to the edges of the DNA bases, allowing them to "read" the DNA sequence. Most of these base-interactions are made in the major groove, where the bases are most accessible.

DNA-modifying Enzymes

Nucleases and Ligases

Nucleases are enzymes that cut DNA strands by catalyzing the hydrolysis of the phosphodiester bonds. Nucleases that hydrolyse nucleotides from the ends of DNA strands are called exonucleases, while endonucleases cut within strands.

The most frequently used nucleases in molecular biology are the restriction endonucleases, which cut DNA at specific sequences. For instance, the EcoRV enzyme shown to the left recognizes the 6-base sequence 52 -GAT | ATC-32 and makes a cut at the vertical line. In nature, these enzymes protect bacteria against phage infection by digesting the phage DNA when it enters the bacterial cell, acting as part of the restriction modification system. In technology, these sequence-specific nucleases are used in molecular cloning and DNA fingerprinting.

Enzymes called DNA ligases can rejoin cut or broken DNA strands. Ligases are particularly important in lagging strand DNA replication, as they join together the short segments of DNA produced at the replication fork into a complete copy of the DNA template. They are also used in DNA repair and genetic recombination.

Topoisomerases and Helicases

Topoisomerases are enzymes with both nuclease and ligase activity. These proteins change the amount of supercoiling in DNA. Some of these enzymes work by cutting the DNA helix and allowing one section to rotate, thereby reducing its level of supercoiling; the enzyme then seals the DNA break. Other types of these enzymes are capable of cutting one DNA helix and then passing a second strand of DNA through this break, before rejoining the helix. Topoisomerases are required for many processes involving DNA, such as DNA replication and transcription.

Helicases are proteins that are a type of molecular motor. They use the chemical energy in nucleoside triphosphates, predominantly ATP, to break hydrogen bonds between bases and unwind the DNA double helix into single strands. These enzymes are essential for most processes where enzymes need to access the DNA bases.

Polymerases

Polymerases are enzymes that synthesize polynucleotide chains from nucleoside triphosphates. The sequence of their products are copies of existing polynucleotide chains - which are called *templates*. These enzymes function by adding nucleotides onto the 32 hydroxyl group of the previous nucleotide in a DNA strand. As a consequence, all polymerases work in a 52 to 32 direction. In the active site of these enzymes, the incoming nucleoside triphosphate base-pairs to the template: this allows polymerases to accurately synthesize the complementary strand of their template. Polymerases are classified according to the type of template that they use.

In DNA replication, a DNA-dependent DNA polymerase makes a copy of a DNA sequence. Accuracy is vital in this process, so many of these polymerases have a proofreading activity. Here, the polymerase recognizes the occasional mistakes in the synthesis reaction by the lack of base pairing between the mismatched nucleotides. If a mismatch is detected, a 32 to 52 exonuclease activity is activated and the incorrect base removed. In most organisms, DNA polymerases function in a large complex called the replisome that contains multiple accessory subunits, such as the DNA clamp or helicases.

RNA-dependent DNA polymerases are a specialized class of polymerases that copy the sequence of an RNA strand into DNA. They include reverse transcriptase, which is a viral enzyme involved in the infection of cells by retroviruses, and telomerase, which is required for the replication of telomeres. Telomerase is an unusual polymerase because it contains its own RNA template as part of its structure.

Transcription is carried out by a DNA-dependent RNA polymerase that copies the sequence of a DNA strand into RNA. To begin transcribing a gene, the RNA polymerase binds to a sequence of DNA called a promoter and separates the DNA strands. It then copies the gene sequence into a messenger RNA transcript until it reaches a region of DNA called the terminator, where it halts and detaches from the DNA. As with human DNA-dependent DNA polymerases, RNA polymerase II, the enzyme that transcribes most of the genes in the human genome, operates as part of a large protein complex with multiple regulatory and accessory subunits.

Genetic Recombination

A DNA helix usually does not interact with other segments of DNA, and in human cells the different chromosomes even occupy separate areas in the nucleus called "chromosome territories". This physical separation of different chromosomes is important for the ability of DNA to function as a stable repository for information, as one of the few times chromosomes interact is during chromosomal crossover when they recombine. Chromosomal crossover is when two DNA helices break, swap a section and then rejoin.

Recombination allows chromosomes to exchange genetic information and produces new combinations of genes, which increases the efficiency of natural selection and can be important in the rapid evolution of new proteins. Genetic recombination can also be involved in DNA repair, particularly in the cell's response to double-strand breaks.

The most common form of chromosomal crossover is homologous recombination, where the two chromosomes involved share very similar sequences. Non-homologous recombination can be damaging to cells, as it can produce chromosomal translocations and genetic abnormalities. The recombination reaction is catalyzed by enzymes known as recombinases, such as RAD51. The first step in recombination is a double-stranded break either caused by an endonuclease or damage to the DNA. A series of steps catalyzed in part by the recombinase then leads to joining of the two helices by at least one Holliday junction, in which a segment of a single strand in each helix is annealed to the complementary strand in the other helix. The Holliday junction is a tetrahedral junction structure that can be moved along the pair of chromosomes, swapping one strand for another. The recombination reaction is then halted by cleavage of the junction and re-ligation of the released DNA.

Evolution

DNA contains the genetic information that allows all modern living things to function, grow and reproduce. However, it is unclear how long in the 4-billion-year history of life DNA has performed this function, as it has been proposed that the earliest forms of life may have used RNA as their genetic material. RNA may have acted as the central part of early cell metabolism as it can both transmit genetic information and carry out catalysis as part of ribozymes.

This ancient RNA world where nucleic acid would have been used for both catalysis and genetics may have influenced the evolution of the current genetic code based on four nucleotide bases. This would occur, since the number of different bases in such an organism is a trade-off between a small number of bases increasing replication accuracy and a large number of bases increasing the catalytic efficiency of ribozymes. However, there is no direct evidence of ancient genetic systems, as recovery of DNA from most fossils is impossible. This is because DNA will survive in the environment for less than one million years and slowly degrades into short fragments in solution. Claims for older DNA have been made, most notably a report of the isolation of a viable bacterium from a salt crystal 250 million years old, but these claims are controversial.

Uses in Technology

Genetic Engineering

Methods have been developed to purify DNA from organisms, such as phenol-chloroform extraction, and to manipulate it in the laboratory, such as restriction digests and the polymerase chain reaction. Modern biology and biochemistry make intensive use of these techniques in recombinant DNA technology. Recombinant DNA is a man-made DNA sequence that has been assembled from other DNA sequences. They can be transformed into organisms in the form of plasmids or in the appropriate format, by using a viral vector. The genetically modified organisms produced can be used to produce products such as recombinant proteins, used in medical research, or be grown in agriculture.

Forensics

Forensic scientists can use DNA in blood, semen, skin, saliva or hair found at a crime scene to identify a matching DNA of an individual, such as a perpetrator. This process is formally termed DNA profiling, but may also be called "genetic fingerprinting". In DNA profiling, the lengths of variable sections of repetitive DNA, such as short tandem repeats and minisatellites, are compared between people.

This method is usually an extremely reliable technique for identifying a matching DNA. However, identification can be complicated if the scene is contaminated with DNA from several people. DNA profiling was developed in 1984 by British geneticist Sir Alec Jeffreys, and first used in forensic science to convict Colin Pitchfork in the 1988 Enderby murders case.

People convicted of certain types of crimes may be required to provide a sample of DNA for a database. This has helped investigators solve old cases where only a DNA sample was obtained from the scene. DNA profiling can also be used to identify victims of mass casualty incidents. On the other hand, many convicted people have been released from prison on the basis of DNA techniques, which were not available when a crime had originally been committed.

Bioinformatics

Bioinformatics involves the manipulation, searching, and data mining of biological data, and this includes DNA sequence data. The development of techniques to store and search DNA sequences have led to widely applied advances in computer science, especially string searching algorithms, machine learning and database theory. String searching or matching algorithms, which find an occurrence of a sequence of letters inside a larger sequence of letters, were developed to search for specific sequences of nucleotides.

The DNA sequenced may be aligned with other DNA sequences to identify homologous sequences and locate the specific mutations that make them distinct. These techniques, especially multiple sequence alignment, are used in studying phylogenetic relationships and protein function.

Data sets representing entire genomes' worth of DNA sequences, such as those produced by the Human Genome Project, are difficult to use without the annotations that identify the locations of genes and regulatory elements on each chromosome.

Regions of DNA sequence that have the characteristic patterns associated with protein- or RNA-coding genes can be identified by gene finding algorithms, which allow researchers to predict the presence of particular gene products and their possible functions in an organism even before they have been isolated experimentally. Entire genomes may also be compared which can shed light on the evolutionary history of particular organism and permit the examination of complex evolutionary events.

DNA Nanotechnology

DNA nanotechnology uses the unique molecular recognition properties of DNA and other nucleic acids to create self-assembling branched DNA complexes with useful properties. DNA is thus used as a structural material rather than as a carrier of biological information. This has led to the creation of two-dimensional periodic lattices (both tile-based as well as using the "DNA origami" method) as well as three-dimensional structures in the shapes of polyhedra. Nanomechanical devices and algorithmic self-assembly have also been demonstrated, and these DNA structures have been used to template the arrangement of other molecules such as gold nanoparticles and streptavidin proteins.

History and Anthropology

Because DNA collects mutations over time, which are then inherited, it contains historical information, and, by comparing DNA sequences, geneticists can infer the evolutionary history of organisms, their phylogeny. This field of phylogenetics is a powerful tool in evolutionary biology. If DNA sequences within a species are compared, population geneticists can learn the history of particular populations. This can be used in studies ranging from ecological genetics to anthropology; For example, DNA evidence is being used to try to identify the Ten Lost Tribes of Israel.

DNA has also been used to look at modern family relationships, such as establishing family relationships between the descendants of Sally Hemings and Thomas Jefferson. This usage is closely related to the use of DNA in criminal investigations detailed above. Indeed, some criminal investigations have been solved when DNA from crime scenes has matched relatives of the guilty individual.

4

Mutation of Molecular Biology

Mutations are changes in a genomic sequence: the DNA sequence of a cell's genome or the DNA or RNA sequence of a virus. Mutations are caused by radiation, viruses, transposons and mutagenic chemicals, as well as errors that occur during meiosis or DNA replication. They can also be induced by the organism itself, by cellular processes such as hypermutation. Mutation can result in several different types of change in DNA sequences; these can either have no effect, alter the product of a gene, or prevent the gene from functioning properly or completely.

Studies in the fly *Drosophila melanogaster* suggest that if a mutation changes a protein produced by a gene, this will probably be harmful, with about 70 percent of these mutations having damaging effects, and the remainder being either neutral or weakly beneficial. Due to the damaging effects that mutations can have on cells, organisms have evolved mechanisms such as DNA repair to remove mutations. Therefore, the optimal mutation rate for a species is a trade-off between costs of a high mutation rate, such as deleterious mutations, and the metabolic costs of maintaining systems to reduce the mutation rate, such as DNA repair enzymes. Viruses that use RNA as their genetic material have rapid mutation rates, which can be an advantage since these viruses will evolve constantly and rapidly, and thus evade the defensive responses of e.g. the human immune system.

Description

Mutations can involve large sections of DNA becoming duplicated, usually through genetic recombination. These duplications are a major source of raw material for evolving new genes, with tens to hundreds of genes duplicated in animal genomes every million years. Most genes belong to larger families of genes of shared ancestry. Novel genes are produced by several methods, commonly through the duplication and mutation of an ancestral gene, or by recombining parts of different genes to form new combinations with new functions. Here, domains act as modules, each with a particular and independent function, that can be mixed together to produce genes encoding new proteins with novel properties. For example, the human eye uses four genes to make structures that sense light: three for colour vision and one for night vision; all four arose from a single ancestral gene. Another advantage of duplicating a gene (or even an entire genome) is that this increases redundancy; this allows one gene in the pair to acquire a new function while the other copy performs the original function. Other types of mutation occasionally create new genes from previously noncoding DNA.

Changes in chromosome number may involve even larger mutations, where segments of the DNA within chromosomes break and then rearrange. For example, two chromosomes in the *Homo* genus fused to produce human chromosome 2; this fusion did not occur in the lineage of the other apes, and they retain these separate chromosomes. In evolution, the most important role of such chromosomal rearrangements may be to accelerate the divergence of a population into new species by making populations less likely to interbreed, and thereby preserving genetic differences between these populations.

Sequences of DNA that can move about the genome, such as transposons, make up a major fraction of the genetic material of plants and animals, and may have been important in the evolution of genomes. For example, more than a million copies of the Alu sequence are present in the human genome, and these sequences have now been recruited to perform functions such as regulating gene expression. Another effect of these mobile DNA sequences is that when they move within a genome, they can mutate or delete existing genes and thereby produce genetic diversity.

In multicellular organisms with dedicated reproductive cells, mutations can be subdivided into germ line mutations, which can be passed on to descendants through their reproductive cells, and somatic mutations (also called acquired mutations), which involve cells outside the dedicated reproductive group and which are not usually transmitted to descendants. If the organism can reproduce asexually through mechanisms such as cuttings or budding the distinction can become blurred.

For example, plants can sometimes transmit somatic mutations to their descendants asexually or sexually where flower buds develop in somatically mutated parts of plants. A new mutation that was not inherited from either parent is called a *de novo* mutation. The source of the mutation is unrelated to the consequence, although the consequences are related to which cells were mutated. Nonlethal mutations accumulate within the gene pool and increase the amount of genetic variation. The abundance of some genetic changes within the gene pool can be reduced by natural selection, while other "more favourable" mutations may accumulate and result in adaptive evolutionary changes. For example, a butterfly may produce offspring with new mutations. The majority of these mutations will have no effect; but one might change the colour of one of the butterfly's offspring, making it harder (or easier) for predators to see. If this colour change is advantageous, the chance of this butterfly surviving and producing its own offspring are a little better, and over time the number of butterflies with this mutation may form a larger percentage of the population.

Neutral mutations are defined as mutations whose effects do not influence the fitness of an individual. These can accumulate over time due to genetic drift. It is believed that the overwhelming majority of mutations have no significant effect on an organism's fitness. Also, DNA repair mechanisms are able to mend most changes before they become permanent mutations, and many organisms have mechanisms for eliminating otherwise permanently mutated somatic cells. Mutation is generally accepted by biologists as the mechanism by which natural selection acts, generating advantageous new traits that survive and multiply in offspring as well as disadvantageous traits, in less fit offspring, that tend to die out.

Causes

Two classes of mutations are spontaneous mutations (molecular decay) and induced mutations caused by mutagens.

Spontaneous mutations on the molecular level can be caused by:

- Tautomerism – A base is changed by the repositioning of a hydrogen atom, altering the hydrogen bonding pattern of that base resulting in incorrect base pairing during replication.
- Depurination – Loss of a purine base (A or G) to form an apurinic site (AP site).
- Deamination – Hydrolysis changes a normal base to an atypical base containing a keto group in place of the original amine group. Examples include C '! U and A '! HX (hypoxanthine), which can be corrected by DNA repair mechanisms; and 5MeC (5-methylcytosine) '! T, which is less likely to be detected as a mutation because thymine is a normal DNA base.
- Slipped strand mispairing-Denaturation of the new strand from the template during replication, followed by renaturation in a different spot ("slipping"). This can lead to insertions or deletions.

Induced mutations on the molecular level can be caused by:

- Chemicals
  - o Hydroxylamine NH_2OH
  - o Base analogs (e.g. BrdU)
  - o Alkylating agents (e.g. *N*-ethyl-*N*-nitrosourea) These agents can mutate both replicating and non-replicating DNA. In contrast, a base analog can only mutate the DNA when the analog is incorporated in replicating the DNA. Each of these classes of chemical mutagens has certain effects that then lead to transitions, transversions, or deletions.
  - o Agents that form DNA adducts (e.g. ochratoxin A metabolites)
  - o DNA intercalating agents (e.g. ethidium bromide)
  - o DNA crosslinkers
  - o Oxidative damage
  - o Nitrous acid converts amine groups on A and C to diazo groups, altering their hydrogen bonding patterns which leads to incorrect base pairing during replication.
- Radiation
- o Ultraviolet radiation (nonionizing radiation). Two nucleotide bases in DNA – cytosine and thymine – are most vulnerable to radiation that can change their properties. UV light can induce adjacent pyrimidine bases in a DNA strand to become covalently joined as a pyrimidine dimer. UV radiation, particularly longer-wave UVA, can also cause oxidative damage to DNA.
- o Ionizing radiation
- Viral infections.

DNA has so-called hotspots, where mutations occur up to 100 times more frequently than the normal mutation rate. A hotspot can be at an unusual base, e.g., 5-methylcytosine. Mutation rates also vary across species. Evolutionary biologists have theorized that higher mutation rates are beneficial in some situations, because they allow organisms to evolve and therefore adapt more quickly to their environments. For example, repeated exposure of bacteria to antibiotics, and selection of resistant mutants, can result in the selection of bacteria that have a much higher mutation rate than the original population (mutator strains).

Classification of Mutation Types

The sequence of a gene can be altered in a number of ways. Gene mutations have varying effects on health depending on where they occur and whether they alter the function of essential proteins. Mutations in the structure of genes can be classified as:

- Small-scale mutations, such as those affecting a small gene in one or a few nucleotides, including:
  - o Point mutations, often caused by chemicals or malfunction of DNA replication, exchange a single nucleotide for another. These changes are classified as transitions or transversions. Most common is the transition that exchanges a purine for a purine (A "! G) or a pyrimidine for a pyrimidine, (C "! T). A transition can be caused by nitrous acid, base mis-pairing, or mutagenic base analogs such as 5-bromo-2-deoxyuridine (BrdU). Less common is a transversion, which exchanges a purine for a pyrimidine or a pyrimidine for a purine (C/T "! A/G). An example of a transversion is adenine (A) being converted into a cytosine (C). A point mutation can be reversed by another point mutation, in which the nucleotide is changed back to its original state (true reversion) or by second-site reversion (a complementary mutation elsewhere that results in regained gene functionality). Point mutations that occur within the protein coding region of a gene may be classified into three kinds, depending upon what the erroneous codon codes for:
    - * Silent mutations: which code for the same amino acid.
    - * Missense mutations: which code for a different amino acid.
    - * Nonsense mutations: which code for a stop and can truncate the protein.
  - o Insertions add one or more extra nucleotides into the DNA. They are usually caused by transposable elements, or errors during replication of repeating elements (e.g. AT repeats). Insertions in the coding region of a gene may alter splicing of the mRNA (splice site mutation), or cause a shift in the reading frame (frameshift), both of which can significantly alter the gene product. Insertions can be reverted by excision of the transposable element.
  - o Deletions remove one or more nucleotides from the DNA. Like insertions, these mutations can alter the reading frame of the gene. They are generally irreversible: though exactly the same sequence might theoretically be restored by an insertion, transposable elements able to revert a very short deletion (say 1–2 bases) in *any* location are either highly unlikely to exist or do not exist at all. Note that a deletion is not the exact opposite of an insertion: the former is quite random while the latter consists of a specific sequence inserting at locations that are not entirely random or even quite narrowly defined.

- Large-scale mutations in chromosomal structure, including:
  - Amplifications (or gene duplications) leading to multiple copies of all chromosomal regions, increasing the dosage of the genes located within them.
  - Deletions of large chromosomal regions, leading to loss of the genes within those regions.
  - Mutations whose effect is to juxtapose previously separate pieces of DNA, potentially bringing together separate genes to form functionally distinct fusion genes (e.g. bcr-abl). These include:
    - Chromosomal translocations: interchange of genetic parts from nonhomologous chromosomes.
    - Interstitial deletions: an intra-chromosomal deletion that removes a segment of DNA from a single chromosome, thereby apposing previously distant genes. For example, cells isolated from a human astrocytoma, a type of brain tumour, were found to have a chromosomal deletion removing sequences between the "fused in glioblastoma" gene and the receptor tyrosine kinase "ros", producing a fusion protein (FIG-ROS). The abnormal FIG-ROS fusion protein has constitutively active kinase activity that causes oncogenic transformation (a transformation from normal cells to cancer cells).
    - Chromosomal inversions: reversing the orientation of a chromosomal segment.
  - Loss of heterozygosity: loss of one allele, either by a deletion or recombination event, in an organism that previously had two different alleles.

By Effect on Function

- Loss-of-function mutations are the result of gene product having less or no function. When the allele has a complete loss of function (null allele) it is often called an amorphic mutation. Phenotypes associated with such mutations are most often recessive. Exceptions are when the organism is haploid, or when the reduced dosage of a normal gene product is not enough for a normal phenotype.
- Gain-of-function mutations change the gene product such that it gains a new and abnormal function. These mutations usually have dominant phenotypes. Often called a neomorphic mutation.
- Dominant negative mutations (also called antimorphic mutations) have an altered gene product that acts antagonistically to the wild-type allele. These mutations usually result in an altered molecular function (often

inactive) and are characterised by a dominant or semi-dominant phenotype. In humans, Marfan syndrome is an example of a dominant negative mutation occurring in an autosomal dominant disease. In this condition, the defective glycoprotein product of the fibrillin gene (FBN1) antagonizes the product of the normal allele.

- Lethal mutations are mutations that lead to the death of the organisms which carry the mutations.
- A back mutation or reversion is a point mutation that restores the original sequence and hence the original phenotype.

By Effect on Fitness

In applied genetics it is usual to speak of mutations as either harmful or beneficial.

- A harmful mutation is a mutation that decreases the fitness of the organism.
- A beneficial mutation is a mutation that increases fitness of the organism, or which promotes traits that are desirable.

In theoretical population genetics, it is more usual to speak of such mutations as deleterious or advantageous. In the neutral theory of molecular evolution, genetic drift is the basis for most variation at the molecular level.

- A neutral mutation has no harmful or beneficial effect on the organism. Such mutations occur at a steady rate, forming the basis for the molecular clock.
- A deleterious mutation has a negative effect on the phenotype, and thus decreases the fitness of the organism.
- An advantageous mutation has a positive effect on the phenotype, and thus increases the fitness of the organism.
- A nearly neutral mutation is a mutation that may be slightly deleterious or advantageous, although most nearly neutral mutations are slightly deleterious.

In reality, viewing the fitness effects of mutations in these discrete categories is an oversimplification. Attempts have been made to infer the distribution of fitness effects using mutagenesis experiments or theoretical models applied to molecular sequence data. However, the current distribution is still uncertain, and some aspects of the distribution likely vary between species.

By Inheritance

- inheritable generic in pro-generic tissue or cells on path to be changed to gametes.
- non inheritable somatic (e.g., carcinogenic mutation)
- non inheritable post mortem aDNA mutation in decaying remains.

By pattern of inheritance The human genome contains two copies of each gene – a paternal and a maternal allele.

- A heterozygous mutation is a mutation of only one allele.
- A homozygous mutation is an identical mutation of both the paternal and maternal alleles.
- Compound heterozygous mutations or a genetic compound comprises two different mutations in the paternal and maternal alleles.
- A wildtype or homozygous non-mutated organism is one in which neither allele is mutated.

By Impact on Protein Sequence

- A frameshift mutation is a mutation caused by insertion or deletion of a number of nucleotides that is not evenly divisible by three from a DNA sequence. Due to the triplet nature of gene expression by codons, the insertion or deletion can disrupt the reading frame, or the grouping of the codons, resulting in a completely different translation from the original. The earlier in the sequence the deletion or insertion occurs, the more altered the protein produced is.
- A nonsense mutation is a point mutation in a sequence of DNA that results in a premature stop codon, or a *nonsense codon* in the transcribed mRNA, and possibly a truncated, and often nonfunctional protein product.
- Missense mutations or *nonsynonymous mutations* are types of point mutations where a single nucleotide is changed to cause substitution of a different amino acid. This in turn can render the resulting protein nonfunctional. Such mutations are responsible for diseases such as Epidermolysis bullosa, sickle-cell disease, and SOD1 mediated ALS.
- A neutral mutation is a mutation that occurs in an amino acid codon which results in the use of a different, but chemically similar, amino acid. The similarity between the two is enough that little or no change is often rendered in the protein. For example, a change from AAA to AGA will encode lysine, a chemically similar molecule to the intended arginine.
- Silent mutations are mutations that do not result in a change to the amino acid sequence of a protein. They may occur in a region that does not code for a protein, or they may occur within a codon in a manner that does not alter the final amino acid sequence. The phrase *silent mutation* is often used interchangeably with the phrase *synonymous mutation*; however, synonymous mutations are a subcategory of the former, occurring only within exons. The name silent could be a

misnomer. For example, a silent mutation in the exon/intron border may lead to alternative splicing by changing the splice site, thereby leading to a changed protein.

Special Classes

Conditional mutation is a mutation that has wild-type (or less severe) phenotype under certain "permissive" environmental conditions and a mutant phenotype under certain "restrictive" conditions. For example, a temperature-sensitive mutation can cause cell death at high temperature (restrictive condition), but might have no deleterious consequences at a lower temperature (permissive condition).

Nomenclature

Nomenclature of mutations specify the type of mutation and base or amino acid changes.

- Nucleotide substitution (e.g. 76A>T)- The number is the position of the nucleotide from the 5' end, the first letter represents the wild type nucleotide, and the second letter represents the nucleotide which replaced the wild type. In the given example, the adenine at the 76th position was replaced by a thymine.
    - If it becomes necessary to differentiate between mutations in genomic DNA, mitochondrial DNA, and RNA, a simple convention is used. For example, if the 100th base of a nucleotide sequence mutated from G to C, then it would be written as g.100G>C if the mutation occurred in genomic DNA, m.100G>C if the mutation occurred in mitochondrial DNA, or r.100g>c if the mutation occurred in RNA. Note that for mutations in RNA, the nucleotide code is written in lower case.
- Amino acid substitution (e.g. D111E) – The first letter is the one letter code of the wild type amino acid, the number is the position of the amino acid from the N terminus, and the second letter is the one letter code of the amino acid present in the mutation. Nonsense mutations are represented with an X for the second amino acid (e.g. D111X).
- Amino acid deletion – The Greek letter Ä (delta) indicates a deletion. The letter refers to the amino acid present in the wild type and the number is the position from the N terminus of the amino acid were it to be present as in the wild type.

Harmful Mutations

Changes in DNA caused by mutation can cause errors in protein sequence, creating partially or completely non-functional proteins. To function correctly, each cell depends on thousands of proteins to function in the right places at the right times. When a mutation alters a protein that plays a critical role in the body, a medical condition can result. A condition caused by mutations in one or more genes is called a genetic disorder.

Some mutations alter a gene's DNA base sequence but do not change the function of the protein made by the gene. Studies of the fly *Drosophila melanogaster* suggest that if a mutation does change a protein, this will probably be harmful, with about 70 percent of these mutations having damaging effects, and the remainder being either neutral or weakly beneficial. However, studies in yeast have shown that only 7% of mutations that are not in genes are harmful.

If a mutation is present in a germ cell, it can give rise to offspring that carries the mutation in all of its cells. This is the case in hereditary diseases. On the other hand, a mutation may occur in a somatic cell of an organism. Such mutations will be present in all descendants of this cell within the same organism, and certain mutations can cause the cell to become malignant, and thus cause cancer.

Often, gene mutations that could cause a genetic disorder are repaired by the DNA repair system of the cell. Each cell has a number of pathways through which enzymes recognize and repair mistakes in DNA. Because DNA can be damaged or mutated in many ways, the process of DNA repair is an important way in which the body protects itself from disease.

Beneficial Mutations

Although most mutations that change protein sequences are neutral or harmful, some mutations have a positive effect on an organism. In this case, the mutation may enable the mutant organism to withstand particular environmental stresses better than wild-type organisms, or reproduce more quickly. In these cases a mutation will tend to become more common in a population through natural selection.

For example, a specific 32 base pair deletion in human CCR5 (CCR5-Ä32) confers HIV resistance to homozygotes and delays AIDS onset in heterozygotes. The CCR5 mutation is more common in those of European descent. One possible explanation of the etiology of the relatively high frequency of CCR5-Ä32 in the European population is that it conferred resistance to the bubonic plague in mid-14th century Europe. People with this mutation were more likely to survive infection; thus its frequency in the population increased. This theory could explain why this mutation is not found in southern Africa, where the bubonic plague never reached. A newer theory suggests that the selective pressure on the CCR5 Delta 32 mutation was caused by smallpox instead of the bubonic plague.

Another example, is Sickle cell disease which is a blood disorder in which the body produces an abnormal type of the oxygen-carrying substance hemoglobin in the red blood cells. One-third of all indigenous inhabitants of Sub-Saharan Africa carry the gene, because in areas where malaria is common, there is a survival value in carrying only a single sickle-cell gene (sickle cell trait). Those with only one of the two alleles of the sickle-cell disease are more resistant to malaria, since the infestation of the malaria plasmodium is halted by the sickling of the cells which it infests.

Prion Mutation

Prions are proteins and do not contain genetic material. However, prion replication has been shown to be subject to mutation and natural selection just like other forms of replication.

Genome

In modern molecular biology and genetics, the genome is the entirety of an organism's hereditary information. It is encoded either in DNA or, for many types of virus, in RNA. The genome includes both the genes and the non-coding sequences of the DNA.

Origin of Term

The term was adapted in 1920 by Hans Winkler, Professor of Botany at the University of Hamburg, Germany. In Greek, the word *genome* means I become, I am born, to come into being. The Oxford English Dictionary suggests the name to be a blend of the words *gene* and *chromosome*. A few related-*ome* words already existed, such as *biome* and *rhizome*, forming a vocabulary into which *genome* fits systematically.

Some organisms have multiple copies of chromosomes, diploid, triploid, tetraploid and so on. In classical genetics, in a sexually reproducing organism (typically eukarya) the gamete has half of the number of chromosome of the somatic cell and the genome is a full set of chromosomes in a gamete. In haploid organisms, including cells of bacteria, archaea, and in organelles including mitochondria and chloroplasts, or viruses, that similarly contain genes, the single or set of circular and/or linear chains of DNA (or RNA for some viruses), likewise constitute the *genome*. The term genome can be applied specifically to mean that stored on a complete set of *nuclear DNA* (i.e., the "nuclear genome") but can also be applied to that stored within organelles that contain their own DNA, as with the "mitochondrial genome" or the "chloroplast genome".

Additionally, the genome can comprise nonchromosomal genetic elements such as viruses, plasmids, and transposable elements. When people say that the genome of a sexually reproducing species has been "sequenced", typically they are referring to a determination of the sequences of one set of autosomes and one of each type of sex chromosome, which together represent both of the possible sexes. Even in species that exist in only one sex, what is described as "a genome sequence" may be a composite read from the chromosomes of various individuals.

In general use, the phrase "genetic makeup" is sometimes used conversationally to mean the genome of a particular individual or organism. The study of the global properties of genomes of related organisms is usually referred to as genomics, which distinguishes it from genetics which generally studies the properties of single genes or groups of genes. Both the number of base pairs and the number of genes vary widely from one species to another, and there is only a rough correlation between the two (an observation known as the C-value paradox). At present, the highest known number of genes is around 60,000, for the protozoan causing trichomoniasis, almost three times as many as in the human genome.

An analogy to the human genome stored on DNA is that of instructions stored in a library:

- The library would contain 46 books (chromosomes)
- The books range in size from 400 to 3340 pages (genes)
- which is 48 to 250 million letters (A,C,G,T) per book.
- Hence the library contains over six billion letters total;
- The library fits into a cell nucleus the size of a pinpoint;
- A copy of the library (all 46 books) is contained in almost every cell of our body.

Types

Most biological entities that are more complex than a virus sometimes or always carry additional genetic material besides that which resides in their chromosomes. In some contexts, such as sequencing the genome of a pathogenic microbe, "genome" is meant to include information stored on this auxiliary material, which is carried in plasmids. In such circumstances then, "genome" describes all of the genes and information on non-coding DNA that have the potential to be present.

In eukaryotes such as plants, protozoa and animals, however, "genome" carries the typical connotation of only information on chromosomal DNA. So although these organisms contain chloroplasts and/or mitochondria that have their own DNA, the genetic information contained by DNA within these organelles is not considered part of the genome. In fact, mitochondria are sometimes said to have their own genome often referred to as the "mitochondrial genome". The DNA found within the chloroplast may be referred to as the "plastome".

Genomes and Genetic Variation

Note that a genome does not capture the genetic diversity or the genetic polymorphism of a species. For example, the human genome sequence in principle could be determined from just half the information on the DNA of one cell from one individual.

To learn what variations in genetic information underlie particular traits or diseases requires comparisons across individuals. This point explains the common usage of "genome" (which parallels a common usage of "gene") to refer not to the information in any particular DNA sequence, but to a whole family of sequences that share a biological context.

Although this concept may seem counter intuitive, it is the same concept that says there is no particular shape that is the shape of a cheetah. Cheetahs vary, and so do the sequences of their genomes. Yet both the individual animals and their sequences share commonalities, so one can learn something about cheetahs and "cheetah-ness" from a single example of either.

Sequencing and Mapping

The Human Genome Project was organized to map and to sequence the human genome. Other genome projects include mouse, rice, the plant *Arabidopsis thaliana*, the puffer fish, bacteria like E. coli, etc. In 1976, Walter Fiers at the University of Ghent (Belgium) was the first to establish the complete nucleotide sequence of a viral RNA-genome (bacteriophage MS2). The first DNA-genome project to be completed was the Phage Ö-X174, with only 5386 base pairs, which was sequenced by Fred Sanger in 1977. The first bacterial genome to be completed was that of Haemophilus influenzae, completed by a team at The Institute for Genomic Research in 1995.

The development of new technologies has dramatically decreased the difficulty and cost of sequencing, and the number of complete genome sequences is rising rapidly. Among many genome database sites, the one maintained by the US National Institutes of Health is inclusive.

These new technologies open up the prospect of personal genome sequencing as an important diagnostic tool. A major step toward that goal was the May 2007 *New York Times* announcement that the full genome of DNA pioneer James D. Watson was deciphered.

Whereas a genome sequence lists the order of every DNA base in a genome, a genome map identifies the landmarks. A genome map is less detailed than a genome sequence and aids in navigating around the genome.

Genome Evolution

Genomes are more than the sum of an organism's genes and have traits that may be measured and studied without reference to the details of any particular genes and their products. Researchers compare traits such as *chromosome number* (karyotype), genome size, gene order, codon usage bias, and GC-content to determine what mechanisms could have produced the great variety of genomes that exist today. Duplications play a major role in shaping the genome. Duplications may range from extension of short tandem repeats, to duplication of a cluster of genes, and all the way to duplications of entire chromosomes or even entire genomes. Such duplications are probably fundamental to the creation of genetic novelty.

Horizontal gene transfer is invoked to explain how there is often extreme similarity between small portions of the genomes of two organisms that are otherwise very distantly related. Horizontal gene transfer seems to be common among many microbes. Also, eukaryotic cells seem to have experienced a transfer of some genetic material from their chloroplast and mitochondrial genomes to their nuclear chromosomes.

Transcription

Protein synthesis is the process in which cells build proteins. The term is sometimes used to refer only to protein translation but more often it refers to a multi-step process, beginning with amino acid synthesis and transcription of nuclear DNA into messenger RNA, which is then used as input to translation.

The cistron DNA is transcribed into a variety of RNA intermediates. The last version is used as a template in synthesis of a polypeptide chain. Proteins can often be synthesized directly from genes by translating mRNA. When a protein needs to be available on short notice or in large quantities, a protein precursor is produced. A proprotein is an inactive protein containing one or more inhibitory peptides that can be activated when the inhibitory sequence is removed by proteolysis during posttranslational modification.

A preprotein is a form that contains a signal sequence (an N-terminal signal peptide) that specifies its insertion into or through membranes; i.e., targets them for secretion. The signal peptide is cleaved off in the endoplasmic reticulum. Preproproteins have both sequences (inhibitory and signal) still present.

For synthesis of protein, a succession of tRNA molecules charged with appropriate amino acids have to be brought together with an mRNA molecule and matched up by base-pairing through their anti-codons with each of its successive codons. The amino acids then have to be linked together to extend the growing protein chain, and the tRNAs, relieved of their burdens, have to be released. This whole complex of processes is carried out by a giant multimolecular machine, the ribosome, formed of two main chains of RNA, called ribosomal RNA (rRNA), and more than 50 different proteins. This molecular juggernaut latches onto the end of an mRNA molecule and then trundles along it, capturing loaded tRNA molecules and stitching together the amino acids they carry to form a new protein chain.

Protein biosynthesis, although very similar, is different for prokaryotes and eukaryotes.

Amino Acid Synthesis

Amino acid synthesis is the set of biochemical processes (metabolic pathways) by which the various amino acids are produced from other compounds. The substrates for these processes are various compounds in the organism's diet or growth media. Not all organisms are able to synthesise all amino acids, for example humans are only able to synthesise 12 of the 20 standard amino acids.

A fundamental problem for biological systems is to obtain nitrogen in an easily usable form. This problem is solved by certain microorganisms capable of reducing the inert Na"N molecule (nitrogen gas) to two molecules of ammonia in one of the most remarkable reactions in biochemistry. Nitrogen in the form of ammonia is the source of nitrogen for all the amino acids. The carbon backbones come from the glycolytic pathway, the pentose phosphate pathway, or the citric acid cycle.

In amino acid production, one encounters an important problem in biosynthesis — namely, stereochemical control. Because all amino acids except glycine are chiral, biosynthetic pathways must generate the correct isomer with high fidelity. In each of the 19 pathways for the generation of chiral amino acids, the stereochemistry at the á-carbon atom is established by a transamination reaction that involves pyridoxal phosphate. Almost all the transaminases that catalyze these reactions descend from a common ancestor, illustrating once again that effective solutions to biochemical problems are retained throughout evolution.

Biosynthetic pathways are often highly regulated such that building blocks are synthesized only when supplies are low. Very often, a high concentration of the final product of a pathway inhibits the activity of enzymes that function early in the pathway. Often present are allosteric enzymes capable of sensing and responding to concentrations of regulatory species. These enzymes are similar in functional properties to aspartate transcarbamoylase and its regulators. Feedback and allosteric mechanisms ensure that all twenty amino acids are maintained in sufficient amounts for protein synthesis and other processes.

Amino acids are synthesized from á-ketoacids, and later transaminated from another aminoacid, usually Glutamate. The enzyme involved in this reaction is an aminotransferase. Glutamate itself is formed by amination of á-ketoglutarate:

Nitrogen Fixation

Microorganisms use ATP and reduced ferredoxin, a powerful reductant, to reduce N2 to NH3. An iron-molybdenum cluster in nitrogenase deftly catalyzes the fixation of N2, a very inert molecule. Higher organisms consume the fixed nitrogen to synthesize amino acids, nucleotides, and other nitrogen-containing biomolecules. The major points of entry of NH4+ into metabolism are glutamine or glutamate.

Amino Acids are Made from Intermediates

Of the basic set of 20 amino acids (not counting selenocysteine) there are 8 that human beings cannot synthesize. In addition, the amino acids arginine, cysteine, glycine, glutamine, histidine, proline, serine and tyrosine are considered conditionally essential, meaning they are not normally required in the diet, but must be supplied exogenously to specific populations that do not synthesize it in adequate amounts. (For example, enough arginine is synthesized by the urea cycle to meet the needs of an adult but perhaps not those of a growing child.) Amino acids that need to be obtained from the diet are called essential amino acids. Nonessential amino acids are produced in the body.

The pathways for the synthesis of nonessential amino acids are quite simple. Glutamate dehydrogenase catalyzes the reductive amination of á-ketoglutarate to glutamate. A transamination reaction takes place in the synthesis of most amino acids. At this step, the chirality of the amino acid is established. Alanine and aspartate are synthesized by the transamination of pyruvate and oxaloacetate, respectively. Glutamine is synthesized from NH4+ and glutamate, and asparagine is synthesized similarly. Proline and arginine are derived from glutamate. Serine, formed from 3-phosphoglycerate, is the precursor of glycine and cysteine. Tyrosine is synthesized by the hydroxylation of phenylalanine, an essential amino acid. The pathways for the biosynthesis of essential amino acids are much more complex than those for the nonessential ones.

Tetrahydrofolate, a carrier of activated one-carbon units, plays an important role in the metabolism of amino acids and nucleotides. This coenzyme carries one-carbon units at three oxidation states, which are interconvertible: most reduced—methyl; intermediate—methylene; and most oxidized—formyl, formimino, and methenyl. The major donor of activated methyl groups is S-adenosylmethionine, which is synthesized by the transfer of an adenosyl group from ATP to the sulfur atom of methionine. S-Adenosylhomocysteine is formed when the activated methyl group is transferred to an acceptor. It is hydrolyzed to adenosine and homocysteine, the latter of which is then methylated to methionine to complete the activated methyl cycle.

Amino Acid Biosynthesis is Regulated by Feedback Inhibition

Most of the pathways of amino acid biosynthesis are regulated by feedback inhibition, in which the committed step is allosterically inhibited by the final product. Branched pathways require extensive interaction among the branches that includes both negative and positive regulation. The regulation of glutamine synthetase from E. coli is a striking demonstration of cumulative feedback inhibition and of control by a cascade of reversible covalent modifications.

Amino Acids are Precursors of Many Biomolecules

Amino acids are precursors of a variety of biomolecules. Glutathione serves as a sulfhydryl buffer and detoxifying agent. Glutathione peroxidase, a selenoenzyme, catalyzes the reduction of hydrogen peroxide and organic peroxides by glutathione. Nitric oxide, a short-lived messenger, is formed from arginine. Porphyrins are synthesized from glycine and succinyl CoA, which condense to give ä-aminolevulinate. Two molecules of this intermediate become linked to form porphobilinogen. Four molecules of porphobilinogen combine to form a linear tetrapyrrole, which cyclizes to uroporphyrinogen III. Oxidation and side-chain modifications lead to the synthesis of protoporphyrin IX, which acquires an iron atom to form heme.

Many organisms have the ability to synthesize only a subset of the amino acids they need. Adult humans, for example, need to obtain 10 of the 20 amino acids from their food.

Transcription

In transcription an mRNA chain is generated, with one strand of the DNA double helix in the genome as template. This strand is called the template strand. Transcription can be divided into 3 stages: Initiation, Elongation and Termination, each regulated by a large number of proteins such as transcription factors and coactivators that ensure the correct gene is transcribed.

The DNA strand is read in the 3' to 5' direction and the mRNA is transcribed in the 5' to 3' direction by the RNA polymerase. Transcription occurs in the cell nucleus, where the DNA is held. The DNA structure of the cell is made up of two helixes made up of sugar and phosphate held together by the bases. The sugar and the phosphate are joined together by covalent bond. The DNA is "unzipped" by the enzyme helicase, leaving the single nucleotide chain open to be copied. RNA polymerase reads the DNA strand from 3 prime (3') end to the 5 prime (5') end, while it synthesizes a single strand of messenger RNA in the 5' to 3' direction. The general RNA structure is very similar to the DNA structure, but in RNA the nucleotide uracil takes the place that thymine occupies in DNA. The single strand of mRNA leaves the nucleus through nuclear pores, and migrates into the cytoplasm.

The first product of transcription differs in prokaryotic cells from that of eukaryotic cells, as in prokaryotic cells the product is mRNA, which needs no post-transcriptional modification, while in eukaryotic cells, the first product is called primary transcript, that needs post-transcriptional modification (capping with 7 methyl guanosine, tailing with a poly A tail) to give hnRNA (heterophil nuclear RNA). hnRNA then undergoes splicing of introns (non coding parts of the gene) via spliceosomes to produce the final mRNA.

Translation

The synthesis of proteins is known as translation. Translation occurs in the cytoplasm, where the ribosomes are located. Ribosomes are made of a small and large subunit that surround the mRNA. In translation, messenger RNA (mRNA) is decoded to produce a specific polypeptide according to the rules specified by the trinucleotide genetic code. This uses an mRNA sequence as a template to guide the synthesis of a chain of amino acids that form a protein. Translation proceeds in four phases: activation, initiation, elongation, and termination (all describing the growth of the amino acid chain, or polypeptide that is the product of translation).

In activation, the correct amino acid (AA) is joined to the correct transfer RNA (tRNA). While this is not technically a step in translation, it is required for translation to proceed. The AA is joined by its carboxyl group to the 3' OH of the tRNA by an ester bond. When the tRNA has an amino acid linked to it, it is termed "charged". Initiation involves the small subunit of the ribosome binding to 5' end of mRNA with the help of initiation factors (IF), other proteins that assist the process. Elongation occurs when the next aminoacyl-tRNA (charged tRNA) in line binds to the ribosome along with GTP and an elongation factor. Termination of the polypeptide happens when the A site of the ribosome faces a stop codon (UAA, UAG, or UGA). When this happens, no tRNA can recognize it, but releasing factor can recognize nonsense codons and causes the release of the polypeptide chain. The capacity of disabling or inhibiting translation in protein biosynthesis is used by antibiotics such as: anisomycin, cycloheximide, chloramphenicol, tetracycline, streptomycin, erythromycin, puromycin etc.

Translation is the process of converting the mRNA codon sequences into an amino acid polypeptide chain.

1. Amino acid activation
2. Initiation-A ribosome attaches to the mRNA and starts to code at the FMet codon (usually AUG, sometimes GUG or UUG).

3. Elongation-tRNA brings the corresponding amino acid (which has an anticodon that identifies the amino acid as the corresponding molecule to a codon) to each codon as the ribosome moves down the mRNA strand.
4. Termination-Reading of the final mRNA codon (aka the STOP codon), which ends the synthesis of the peptide chain and releases it.

Events Following Protein Translation

The events following biosynthesis include post-translational modification and protein folding. During and after synthesis, polypeptide chains often fold to assume, so called, native secondary and tertiary structures. This is known as *protein folding*. Many proteins undergo *post-translational modification*. This may include the formation of disulfide bridges or attachment of any of a number of biochemical functional groups, such as acetate, phosphate, various lipids and carbohydrates. Enzymes may also remove one or more amino acids from the leading (amino) end of the polypeptide chain, leaving a protein consisting of two polypeptide chains connected by disulfide bonds.

In general, protein molecules are believed to be modified by small chemical groups, post-translationally. Chemical modifications such as, phosphorylation of serine/threonine, acetylation or methylation of lysine, hydroxylation of proline/lysine, formylation of glycine, glycosylation of serine/threonine/ asparagine, acylation of cysteine, myristoylation of glycine, biotinylation of lysine, ubiquitination, etc. on proteins is a very important issue in relation to properly understanding the biological functions of a given protein.

These much-studied post-translational modifications have become well-established with the discovery of the respective enzymes (kinases for phosphorylation, acetylases for acetylation, methyl transferases for methylation, etc.), which carry on the chemical modifications on the specific amino acid residues. All of these modifications are still believed to have happened as post-translational events. There is no study yet on when actually one particular modification occurs on a given amino acid residue in a given protein. Does it happen when the protein is already formed, or when the amino acid chain is being synthesized, or before the translation of the primary chain has begun?

Since these chemical modifications are related to the biological functions of a protein, it is easy to think that these chemical modifications have happened to the whole protein molecule, after the protein primary chain is fully synthesized; but, if that is the case, we have to consider the fact that the primary chains get folded instantly, (in a similar way as the newly synthesized DNA strands form helixes), to attain its compact-globular conformation ; As most of the primary chains are fairly long [a 5Kd protein may have 40-45 amino acid residues in its primary chain], it is likely that the newly formed amino acid chain tries to remain intact by folding, thereby avoiding its breakdown via lots of proteases present within the cytoplasm. And, no capping event to protect the N-terminal end of the primary sequence (similar to 5' m-RNA capping to protect m-RNAs) is ever discovered for protein primary structure.

So, by folding mechanism, the primary chain, perhaps, avoids the protease attacks. However, once it gets folded, it may be very difficult for the respective enzyme molecule to find out the particular aa residue from the complexity of that compactly folded conformation. In addition, it can be clearly imagined that this enzymatic modification/reaction on a given amino acid requires presence and association of the appropriate enzyme, necessary cofactors, etc. This association is much easier to occur when the amino acid residues in the primary structure are readily available for binding; in other words, it is much difficult for the enzyme molecules to find and to bind to its substrate amino acid residue in a mature protein molecule after its three-dimensional conformation been attained.

So one can think that the modifications can happen while the primary chain of the protein is being synthesized during the translation process on the m-RNA strand; the amino acid residues on the primary chain can be modified instantly and enzymatically by kinases, acetylases, hydroxylases, methyltransferases, etc. to initiate proper folding for protection in order to avoid degradation by proteases, thereby gaining the globular form.

Also, the reader can imagine another scenario in which, while the free amino acid molecules are formed within a cell and become available, they [in that free state] may be modified enzymatically before taking the ride to the translational event; this means that, while the primary chain is being synthesized, the pre-modified amino acid molecules are ready to be engaged. This way, as the primary chain is getting synthesized, it does not have to be modified by the modifying enzymes anymore, and can fold itself instantaneously without concern of degradation. This also arises new thoughts that (i) phosphate, acetyl, methyl, biotinyl, acyl, etc. groups may have the ability to inhibit protease actions on the primary chains; (ii) most or all of the amino acid residues get modified by small chemical groups (so far, only some chemical groups are known).

It is still not fully known exactly when and how the actual modification of a given amino acid residue occurs, at which stage of synthesis, within a protein molecule.

Once the chemically-modified, protease-insensitive, intact protein molecule is generated, it must perform its biological function, which requires its being activated. The activation is probably done by a second set of enzymatic reactions when these chemical groups are removed from the aa residues (or added back). So, the second type of post-translational modifications are the opposite reactions of the above-described type, which are dephosphorylation by phosphatases or phosphoryl transferases, deacetylation by deacetylases or acetyltransferases, demethylation by demethylases or methyl-transferases, ubiquitination, SuMoylation, glycosylation, biotinylation, etc. These are taking place on the whole protein molecule toward generating its activated form to execute a particular function or toward its deactivation followed by its total degradation. As the chemical groups are removed (or added), it is quite clear that the proteins can go through different states of structural/conformational change. Thus, after translation, a protein can change its conformation dynamically while the chemical groups are removed or added enzymatically; these conformational changes in the protein structure help the protein to proceed through its lifecycle, until it is ubiquitinated for its total degradation.

Protein synthesis is important because it is through the use of genetic information to synthesize proteins that the genes in the DNA are expressed in the phenotype. Make sure you understand what proteins are and why they are important. Protein synthesis involves two big steps and several other smaller steps depending on the organism. The two big steps are called transcription and translation. Transcription involves the synthesis of an RNA transcript using the cell's DNA as a template, leading to messenger RNA(mRNA). In eukaryotic cells transcription happens in the nucleus. Once the messenger RNA is made it then is transported to the cytoplasm where the information contained in the mRNA is translated into a sequence of amino acids making up a polypeptide. We first examine the simplest way of looking at protein synthesis as expressed in the so called Central Dogma of Biology, namely that the direction of information flow in the cell is from DNA to mRNA to proteins.

Transcription

Transcription relies on the principle of complementarity to make the RNA transcript. The RNA's are nucleic acids as are the DNA's. The big differences are that RNA is not double sided, RNAs have the nitrogen base uracil in place of thymine and the five carbon sugar in the nucleotides is ribose, not deoxyribose. This process involves several enzymes. Chief among them RNA polymerase. The developing RNA strand is on the right in this diagram. Once the RNA transcript is made, if it is going to be used as messenger RNA is may be edited before it is translated. It's important to realize that all types of RNA are made by transcription but only messenger RNA has the sequence of information that will later be translated into the sequence of amino acids that will become the polypeptide. Just as DNA replication involves a series of enzymes, so to does transcription. The main enzyme is called RNA polymerase. The oval shape under the side of the DNA molecule shown represents a the RNA polymerase. Only one side of the DNA molecule needs to be shown, because typically the other side is not transcribed, at least within the region of the same gene. Once the RNA transcript that will become the messenger RNA is made, it is carried to the cytoplasm of the cell where it translated into, if you will, the language of proteins, namely a sequence of amino acids.

Gene Concept, Gene Bank and Gene Library

Cell division is the process by which a *parent cell* divides into two or more *daughter cells.* Cell division is usually a small segment of a larger cell cycle. This type of cell division in eukaryotes is known as mitosis, and leaves the daughter cell capable of dividing again. The corresponding sort of cell division in prokaryotes is known as binary fission. In another type of cell division present only in eukaryotes, called meiosis, a cell is permanently transformed into a gamete and cannot divide again until fertilization. Right before the parent cell splits, it undergoes DNA replication. For simple unicellular organisms such as the amoeba, one cell division is equivalent to reproduction— an entire new organism is created. On a larger scale, mitotic cell division can create progeny from multicellular organisms, such as plants that grow from cuttings. Cell division also enables asexually reproducing organisms to develop from the one-celled zygote, which itself was produced by cell division from gametes. And after growth, cell division allows for continual construction and repair of the organism. A human being's body experiences about 10,000 trillion cell divisions in a lifetime. The primary concern of cell division is the maintenance of the original cell's genome. Before division can occur, the genomic information which is stored in chromosomes must be replicated, and the duplicated genome separated cleanly between cells. A great deal of cellular infrastructure is involved in keeping genomic information consistent between "generations".

Variants

Cells are classified into two categories: simple, non-nucleated prokaryotic cells, and complex, nucleated eukaryotic cells. By dint of their structural differences, eukaryotic and prokaryotic cells do not divide in the same way.

Furthermore, the pattern of cell division that transforms eukaryotic stem cells into gametes (sperm in males or ova in females) is different from that of eukaryotic somatic (non-germ) cells.

Degradation

Multicellular organisms replace worn-out cells through cell division. In some animals, however, cell division eventually halts. In humans this occurs on average, after 52 divisions, known as the Hayflick limit. The cell is then referred to as senescent. Cells stop dividing because the telomeres, protective bits of DNA on the end of a chromosome required for replication, shorten with each copy, eventually being consumed, as described in the article on telomere shortening. Cancer cells, on the other hand, are not thought to degrade in this way, if at all. An enzyme called telomerase, present in large quantities in cancerous cells, rebuilds the telomeres, allowing division to continue indefinitely.

Mitosis

Mitosis is the process by which a eukaryotic cell separates the chromosomes in its cell nucleus into two identical sets in two nuclei. It is generally followed immediately by cytokinesis, which divides the nuclei, cytoplasm, organelles and cell membrane into two cells containing roughly equal shares of these cellular components. Mitosis and cytokinesis together define the mitotic (M) phase of the cell cycle-the division of the mother cell into two daughter cells, genetically identical to each other and to their parent cell. This accounts for approximately 10% of the cell cycle.

Mitosis occurs exclusively in eukaryotic cells, but occurs in different ways in different species. For example, animals undergo an "open" mitosis, where the nuclear envelope breaks down before the chromosomes separate, while fungi such as *Aspergillus nidulans* and *Saccharomyces cerevisiae* (yeast) undergo a "closed" mitosis, where chromosomes divide within an intact cell nucleus. Prokaryotic cells, which lack a nucleus, divide by a process called binary fission.

The process of mitosis is complex and highly regulated. The sequence of events is divided into phases, corresponding to the completion of one set of activities and the start of the next. These stages are interphase, prophase, prometaphase, metaphase, anaphase and telophase. During the process of mitosis the pairs of chromosomes condense and attach to fibers that pull the sister chromatids to opposite sides of the cell. The cell then divides in cytokinesis, to produce two identical daughter cells.

Because cytokinesis usually occurs in conjunction with mitosis, "mitosis" is often used interchangeably with "mitotic phase". However, there are many cells where mitosis and cytokinesis occur separately, forming single cells with multiple nuclei. This occurs most notably among the fungi and slime moulds, but is found in various different groups. Even in animals, cytokinesis and mitosis may occur independently, for instance during certain stages of fruit fly embryonic development. Errors in mitosis can either kill a cell through apoptosis or cause mutations that may lead to cancer.

The primary result of mitosis is the division of the parent cell's genome into two daughter cells. The genome is composed of a number of chromosomes-complexes of tightly-coiled DNA that contain genetic information vital for proper cell function. Because each resultant daughter cell should be genetically identical to the parent cell, the parent cell must make a copy of each chromosome before mitosis. This occurs during S phase, in interphase, the period that precedes the mitotic phase in the cell cycle where preparation for mitosis occurs.

Each new chromosome now contains two identical copies of itself, called *sister chromatids*, attached together in a specialized region of the chromosome known as the *centromere*. Each sister chromatid is not considered a chromosome in itself, and a chromosome always contains two sister chromatids.

In most eukaryotes, the nuclear envelope that separates the DNA from the cytoplasm disassembles. The chromosomes align themselves in a line spanning the cell. Microtubules, essentially miniature strings, splay out from opposite ends of the cell and shorten, pulling apart the sister chromatids of each chromosome. As a matter of convention, each sister chromatid is now considered a chromosome, so they are renamed to *sister chromosomes*. As the cell elongates, corresponding sister chromosomes are pulled toward opposite ends. A new nuclear envelope forms around the separated sister chromosomes. As mitosis completes cytokinesis is well underway. In animal cells, the cell pinches inward where the imaginary line used to be (the pinching of the cell membrane to form the two daughter cells is called the cleavage furrow), separating the two developing nuclei. In plant cells, the daughter cells will construct a new dividing cell wall between each other. Eventually, the mother cell will be split in half, giving rise to two daughter cells, each with an equivalent and complete copy of the original genome. Prokaryotic cells undergo a process similar to mitosis called binary fission. However, prokaryotes cannot be properly said to undergo mitosis because they lack a nucleus and only have a single chromosome with no centromere.

Phases of Cell Cycle and Mitosis

The mitotic phase is a relatively short period of the cell cycle. It alternates with the much longer *interphase*, where the cell prepares itself for cell division. Interphase is therefore not part of mitosis. Interphase is divided into three phases, G_1, S (synthesis), and G_2 (second gap). During all three phases, the cell grows by producing proteins and cytoplasmic organelles. However, chromosomes are replicated only during the S phase. Thus, a cell grows (G_1), continues to grow as it duplicates its chromosomes (S), grows more and prepares for mitosis (G_2), and finally it divides (M) before restarting the cycle.

Preprophase

In plant cells only, prophase is preceded by a pre-prophase stage. In highly vacuolated plant cells, the nucleus has to migrate into the centre of the cell before mitosis can begin. This is achieved through the formation of a phragmosome, a transverse sheet of cytoplasm that bisects the cell along the future plane of cell division. In addition to phragmosome formation, preprophase is characterized by the formation of a ring of microtubules and actin filaments (called preprophase band) underneath the plasma membrane around the equatorial plane of the future mitotic spindle. This band marks the position where the cell will eventually divide. The cells of higher plants (such as the flowering plants) lack centrioles: with microtubules forming a spindle on the surface of the nucleus and then being organized into a spindle by the chromosomes themselves, after the nuclear membrane breaks down. The preprophase band disappears during nuclear envelope disassembly and spindle formation in prometaphase.

Prophase

Normally, the genetic material in the nucleus is in a loosely bundled coil called chromatin. At the onset of prophase, chromatin condenses together into a highly ordered structure called a chromosome. Since the genetic material has already been duplicated earlier in S phase, the replicated chromosomes have two sister chromatids, bound together at the centromere by the cohesion complex. Chromosomes are visible at high magnification through a light microscope.

Close to the nucleus are structures called centrosomes, which are made of a pair of centrioles. The centrosome is the coordinating centre for the cell's microtubules. A cell inherits a single centrosome at cell division, which replicates before a new mitosis begins, giving a pair of centrosomes. The two centrosomes nucleate microtubules (which may be thought of as cellular ropes or poles) to form the spindle by polymerizing soluble tubulin. Molecular motor proteins then push the centrosomes along these microtubules to opposite sides of the cell. Although centrioles help organize microtubule assembly, they are not essential for the formation of the spindle, since they are absent from plants, and centrosomes are not always used in meiosis.

Prometaphase

The nuclear envelope disassembles and microtubules invade the nuclear space. This is called open mitosis, and it occurs in most multicellular organisms. Fungi and some protists, such as algae or trichomonads, undergo a variation called closed mitosis where the spindle forms inside the nucleus or its microtubules are able to penetrate an intact nuclear envelope.

Each chromosome forms two kinetochores at the centromere, one attached at each chromatid. A kinetochore is a complex protein structure that is analogous to a ring for the microtubule hook; it is the point where microtubules attach themselves to the chromosome. Although the kinetochore structure and function are not fully understood, it is known that it contains some form of molecular motor. When a microtubule connects with the kinetochore, the motor activates, using energy from ATP to "crawl" up the tube toward the originating centrosome. This motor activity, coupled with polymerisation and depolymerisation of microtubules, provides the pulling force necessary to later separate the chromosome's two chromatids.

When the spindle grows to sufficient length, *kinetochore microtubules* begin searching for kinetochores to attach to. A number of *nonkinetochore microtubules* find and interact with corresponding nonkinetochore microtubules from the opposite centrosome to form the mitotic spindle. Prometaphase is sometimes considered part of prophase. In the fishing pole analogy, the kinetochore would be the "hook" that catches a sister chromatid or "fish". The centrosome acts as the "reel" that draws in the spindle fibers or "fishing line".

Metaphase

As microtubules find and attach to kinetochores in prometaphase, the centromeres of the chromosomes convene along the *metaphase plate* or *equatorial plane*, an imaginary line that is equidistant from the two centrosome poles. This even alignment is due to the counterbalance of the pulling powers generated by the opposing kinetochores, analogous to a tug-of-war between people of equal strength. In certain types of cells, chromosomes do not line up at the metaphase plate and instead move back and forth between the poles randomly, only roughly lining up along the midline. Metaphase comes from the Greek *ìåôá* meaning "after."

Because proper chromosome separation requires that every kinetochore be attached to a bundle of microtubules (spindle fibres), it is thought that unattached kinetochores generate a signal to prevent premature progression to anaphase without all chromosomes being aligned. The signal creates the *mitotic spindle checkpoint.*

Anaphase

When every kinetochore is attached to a cluster of microtubules and the chromosomes have lined up along the metaphase plate, the cell proceeds to anaphase. Two events then occur; First, the proteins that bind sister chromatids together are cleaved, allowing them to separate. These sister chromatids, which have now become distinct sister chromosomes, are pulled apart by shortening kinetochore microtubules and move toward the respective centrosomes to which they are attached. Next, the nonkinetochore microtubules elongate, pulling the centrosomes (and the set of chromosomes to which they are attached) apart to opposite ends of the cell. The force that causes the centrosomes to move towards the ends of the cell is still unknown, although there is a theory that suggests that the rapid assembly and breakdown of microtubules may cause this movement. These two stages are sometimes called early and late anaphase. Early anaphase is usually defined as the separation of the sister chromatids, while late anaphase is the elongation of the microtubules and the chromosomes being pulled farther apart. At the end of anaphase, the cell has succeeded in separating identical copies of the genetic material into two distinct populations.

Telophase

Telophase is a reversal of prophase and prometaphase events. It "cleans up" the after effects of mitosis. At telophase, the nonkinetochore microtubules continue to lengthen, elongating the cell even more. Corresponding sister chromosomes attach at opposite ends of the cell. A new nuclear envelope, using fragments of the parent cell's nuclear membrane, forms around each set of separated sister chromosomes. Both sets of chromosomes, now surrounded by new nuclei, unfold back into chromatin. Mitosis is complete, but cell division is not yet complete.

Cytokinesis

Cytokinesis is often mistakenly thought to be the final part of telophase; however, cytokinesis is a separatc process that begins at the same time as telophase. Cytokinesis is technically not even a phase of mitosis, but rather a separate process, necessary for completing cell division. In animal cells, a cleavage furrow (pinch) containing a contractile ring develops where the metaphase plate used to be, pinching off the separated nuclei.

In both animal and plant cells, cell division is also driven by vesicles derived from the Golgi apparatus, which move along microtubules to the middle of the cell. In plants this structure coalesces into a cell plate at the centre of the phragmoplast and develops into a cell wall, separating the two nuclei. The phragmoplast is a microtubule structure typical for higher plants, whereas some green algae use a phycoplast microtubule array during cytokinesis. Each daughter cell has a complete copy of the genome of its parent cell. The end of cytokinesis marks the end of the M-phase.

Significance

Mitosis is important for the maintenance of the chromosomal set; each cell formed receives chromosomes that are alike in composition and equal in number to the chromosomes of the parent cell. Transcription is generally believed to cease during mitosis, but epigenetic mechanisms such as bookmarking function during this stage of the cell cycle to ensure that the "memory" of which genes were active prior to entry into mitosis are transmitted to the daughter cells.

Consequences of Errors

Although errors in mitosis are rare, the process may go wrong, especially during early cellular divisions in the zygote. Mitotic errors can be especially dangerous to the organism because future offspring from this parent cell will carry the same disorder. In *non-disjunction*, a chromosome may fail to separate during anaphase. One daughter cell will receive both sister chromosomes and the other will receive none. This results in the former cell having three chromosomes containing the same genes (two sisters and a homologue), a condition known as *trisomy*, and the latter cell having only one chromosome (the homologous chromosome), a condition known as *monosomy*. These cells are considered aneuploid, a condition often associated with cancer. Mitosis is a traumatic process. The cell goes through dramatic changes in ultrastructure, its organelles disintegrate and reform in a matter of hours, and chromosomes are jostled constantly by probing microtubules. Occasionally, chromosomes may become damaged. An arm of the chromosome may be broken and the fragment lost, causing deletion. The fragment may incorrectly reattach to another, non-homologous chromosome, causing translocation. It may reattach to the original chromosome, but in reverse orientation, causing inversion. Or, it may be treated erroneously as a separate chromosome, causing chromosomal duplication. The effect of these genetic abnormalities depends on the specific nature of the error. It may range from no noticeable effect to cancer induction or organism death.

Endomitosis

Endomitosis is a variant of mitosis without nuclear or cellular division, resulting in cells with many copies of the same chromosome occupying a single nucleus. This process may also be referred to as endoreduplication and the cells as endoploid. An example of a cell that goes through endomitosis is the megakaryocyte.

Binary Fission

Binary fission, or prokaryotic fission, is the form of asexual reproduction and cell division used by all prokaryotes, some protozoa, and some organelles within eukaryotic organisms. This process results in the reproduction of a living prokaryotic cell by division into two parts which each have the potential to grow to the size of the original cell. Mitosis and cytokinesis are not the same as binary fission; specifically, binary fission cannot be divided into prophase, metaphase, anaphase, and telophase because prokaryotes have no nucleus and no centromeres. The ability of some multicellular animals, such as echinoderms and flatworms, to regenerate two whole organisms after having been cut in half, is also not the same as binary fission. Neither is vegetative reproduction of plants.

Process

Binary fission begins with DNA replication. DNA replication starts from an origin of replication, which opens up into a replication bubble (note: prokaryotic DNA replication usually has only 1 origin of replication, whereas eukaryotes have multiple origins of replication). The replication bubble separates the DNA double strand, each strand acts as template for synthesis of a daughter strand by semiconservative replication, until the entire prokaryotic DNA is duplicated. After this replicational process, cell growth occurs. Each circular DNA strand then attaches to the cell membrane.

The cell elongates, causing the two chromosomes to separate. Cell division in bacteria is controlled by the FtsZ, a collection of about a dozen proteins that collect around the site of division. There, they direct assembly of the division septum. The cell wall and plasma membrane starts growing transversely from near the middle of the dividing cell. This separates the parent cell into two nearly equal daughter cells, each having a nuclear body. The cell membrane then invaginates (grows inwards) and splits the cell into two daughter cells, separated by a newly grown cell plate.

Use by Eukaryotic Organelles

Eukaryotic organelles such as mitochondria, chloroplasts, and peroxisomes also reproduce within the eukaryotic cell by binary fission. How they are allotted to one descendant cell or the other during mitosis and cytokinesis is not yet clear.

Gene Bank

Gene banks help preserve genetic material, be it plant or animal. In plants, this could be by freezing cuts from the plant, or stocking the seeds. In animals, this is the freezing of sperm and eggs in zoological freezers until further need. With corals, fragments are taken which are stored in water tanks under controlled conditions. In plants, it is possible to unfreeze the material and propagate it, however, in animals, a living female is required for artificial insemination. While it is often difficult to utilize frozen animal sperm and eggs, there are many examples of it being done successfully. In an effort to conserve agricultural biodiversity, gene banks are used to store and conserve the plant genetic resources of major crop plants and their crop wild relatives. There are many gene banks all over the world, with the Svalbard Global Seed Vault being probably the most famous one.

Genomic Library

A genomic library is a population of host bacteria, each of which carries a DNA molecule that was inserted into a cloning vector, such that the collection of cloned DNA molecules represents the entire genome of the source organism.

This term also represents the collection of all of the vector molecules, each carrying a piece of the chromosomal DNA of the organism, prior to the insertion of these molecules into the host cells.

Types of Vectors

1. Plasmids: Have a capacity of 15 kb.
2. Phage (lambda)s: Have a capacity of 25 kb.
3. Cosmids or Fosmids: Have a capacity of 35-45 kb.
4. Bacterial artificial chromosomes (BAC)(P-1 derived): Have a capacity of 50-300 kb.
5. Yeast artificial chromosomes (YAC): Have a capacity of 300->1500 kb.
6. Human artificial chromosomes (HAC): Have a capacity of >2000 kb.

Creating a Library

The DNA molecules of an organism of interest are isolated. The DNA molecules are then partially digested by an endonuclease restriction enzyme. Sometimes, the DNA molecules are digested for different lengths of time in order to ensure that all the DNA has been digested to manageable sizes. The digested DNA molecules are separated by size using agarose electrophoresis, and a suitable range of lengths of DNA pieces are isolated and ligated into vectors. The vectors can then be taken up by suitable hosts.

The hosts are kept in liquid media and can be frozen at-80°C for a long period of time. Usually the hosts are bacteria that do not contain any plasmids, so as to be sensitive to antibiotics. The process of subdividing genomic DNA into clonable elements and inserting them into hosts is called creating a library, a clone bank or a gene bank. A complete library of host cells will contain all of the genomic DNA of the source organism.

5

Molecular Biology Applications in Clonning

Molecular cloning refers to a set of experimental methods in molecular biology that are used to assemble recombinant DNA molecules and to direct their replication within host organisms. The use of the word *cloning* refers to the fact that the method involves the replication of a single DNA molecule starting from a single living cell to generate a large population of cells containing identical DNA molecules. Molecular cloning generally uses DNA sequences from two different organisms: the species that is the source of the DNA to be cloned, and the species that will serve as the living host for replication of the recombinant DNA. Molecular cloning methods are central to many contemporary areas of modern biology and medicine.

In a conventional molecular cloning experiment, the DNA to be cloned is obtained from an organism of interest, then treated with enzymes in the test tube to generate smaller DNA fragments. Subsequently, these fragments are then combined with vector DNA to generate recombinant DNA molecules.

The recombinant DNA is then introduced into a host organism (typically an easy-to-grow, benign, laboratory strain of *E. coli* bacteria). This will generate a population of organisms in which recombinant DNA molecules are replicated along with the host DNA. Because they contain foreign DNA fragments, these are transgenic or genetically-modified microorganisms (GMO). This process takes advantage of the fact that a single bacterial cell can be induced to take up and replicate a single recombinant DNA molecule. This single cell can then be expanded exponentially to generate a large amount of bacteria, each of which contain copies of the original recombinant molecule. Thus, both the resulting bacterial population, and the recombinant DNA molecule, are commonly referred to as "clones". Strictly speaking, *recombinant DNA* refers to DNA molecules, while *molecular cloning* refers to the experimental methods used to assemble them.

History of Molecular Cloning

Prior to the 1970s, our understanding of genetics and molecular biology was severely hampered by an inability to isolate and study individual genes from complex organisms. This changed dramatically with the advent of molecular cloning methods. Microbiologists, seeking to understand the molecular mechanisms through which bacteria restricted the growth of bacteriophage, isolated restriction endonucleases, enzymes that could cleave DNA molecules only when specific DNA sequences were encountered. They showed that restriction enzymes cleaved chromosome-length DNA molecules at specific locations, and that specific sections of the larger molecule could be purified by size fractionation. Using a second enzyme, DNA ligase, fragments generated by restriction enzymes could be joined in new combinations, termed recombinant DNA. By recombining DNA segments of interest with vector DNA, such as bacteriophage or plasmids, which naturally replicate inside bacteria, large quantities of purified recombinant DNA molecules could be produced in bacterial cultures. The first recombinant DNA molecules were generated and studied in 1972.

Overview

Molecular cloning takes advantage of the fact that the chemical structure of DNA is fundamentally the same in all living organisms. Therefore, if any segment of DNA from any organism is inserted into a DNA segment containing the molecular sequences required for DNA replication, and the resulting recombinant DNA is introduced into the organism from which the replication sequences were obtained, then the foreign DNA will be replicated along with the host cell's DNA in the transgenic organism.

Molecular cloning is similar to polymerase chain reaction (PCR) in that it permits the replication of a specific DNA sequence. The fundamental difference between the two methods is that molecular cloning involves replication of the DNA in a living microorganism, while PCR replicates DNA in an *in vitro* solution, free of living cells.

Steps in Molecular Cloning

In standard molecular cloning experiments, the cloning of any DNA fragment essentially involves seven steps: (1) Choice of host organism and cloning vector, (2) Preparation of vector DNA, (3) Preparation of DNA to be cloned, (4) Creation of recombinant DNA, (5) Introduction of recombinant DNA into host organism, (6) Selection of organisms containing recombinant DNA, (7) Screening for clones with desired DNA inserts and biological properties.

Choice of Host Organism and Cloning Vector

Although a very large number of host organisms and molecular cloning vectors are in use, the great majority of molecular cloning experiments begin with a laboratory strain of the bacterium *E. coli* (Escherichia coli) and a plasmid cloning vector.

E. coli and plasmid vectors are in common use because they are technically sophisticated, versatile, widely available, and offer rapid growth of recombinant organisms with minimal equipment. If the DNA to be cloned is exceptionally large (hundreds of thousands to millions of base pairs), then a bacterial artificial chromosome or yeast artificial chromosome vector is often chosen.

Specialised applications may call for specialised host-vector systems. For example, if the experimentalists wish to harvest a particular protein from the recombinant organism, then an expression vector is chosen that contains appropriate signals for transcription and translation in the desired host organism. Alternatively, if replication of the DNA in different species is desired (for example transfer of DNA from bacteria to plants), then a multiple host range vector (also termed shuttle vector) may be selected.

In practice, however, specialised molecular cloning experiments usually begin with cloning into a bacterial plasmid, followed by subcloning into a specialised vector.

Whatever combination of host and vector are used, the vector almost always contains four DNA segments that are critically important to its function and experimental utility—(1) an origin of DNA replication is necessary for the vector (and recombinant sequences linked to it) to replicate inside the host organism, (2) one or more unique restriction endonuclease recognition sites that serves as sites where foreign DNA may be introduced, (3) a selectable genetic marker gene that can be used to enable the survival of cells that have taken up vector sequences, and (4) an additional gene that can be used for screening which cells contain foreign DNA.

Preparation of Vector DNA

The cloning vector is treated with a restriction endonuclease to cleave the DNA at the site where foreign DNA will be inserted. The restriction enzyme is chosen to generate a configuration at the cleavage site that is compatible with that at the ends of the foreign DNA. Typically, this is done by cleaving the vector DNA and foreign DNA with the same restriction enzyme, for example EcoRI.

Most modern vectors contain a variety of convenient cleavage sites that are unique within the vector molecule (so that the vector can only be cleaved at a single site) and is located within a gene (frequently beta-galactosidase) whose inactivation can be used to distinguish recombinant from non-recombinant organisms at a later step in the process.

To improve the ratio of recombinant to non-recombinant organisms, the cleaved vector may be treated with an enzyme (alkaline phosphatase) that modifies the vector ends in such a way that it cannot replicate within cells unless it contains foreign DNA.

Preparation of DNA to be Cloned

For cloning of genomic DNA, the DNA to be cloned is extracted from the organism of interest. Virtually any tissue source can be used (even tissues from extinct animals), as long as the DNA is not extensively degraded. The DNA is then purified using simple methods to remove contaminating proteins (extraction with phenol), RNA (ribonuclease) and smaller molecules (precipitation and/or chromatography). Polymerase chain reaction (PCR) methods are often used for amplification of specific DNA or RNA (RT-PCR) sequences prior to molecular cloning.

DNA for cloning experiments may also be obtained from RNA using reverse transcriptase (complementary DNA or cDNA cloning), or in the form of synthetic DNA (artificial gene synthesis). cDNA cloning is usually used to obtain clones representative of the mRNA population of the cells of interest, while synthetic DNA is used to obtain any precise sequence defined by the designer.

The purified DNA is then treated with a restriction enzyme to generate fragments with ends capable of being linked to those of the vector. If necessary, short double-stranded segments of DNA containing desired restriction sites may be added to create end structures that are compatible with the vector.

Creation of Recombinant DNA with DNA Ligase

The creation of recombinant DNA is in many ways the simplest step of the molecular cloning process. DNA prepared from the vector and foreign source are simply mixed together at appropriate concentrations and exposed to an enzyme (DNA ligase) that covalently links the ends together.

This joining reaction is often termed ligation. The resulting DNA mixture containing randomly joined ends is then ready for introduction into the host organism. DNA ligase only recognises and acts on the ends of linear DNA molecules, usually resulting a complex mixture of DNA molecules with randomly joined ends. The desired products (vector DNA covalently linked to foreign DNA) will be present, but other sequences (e.g. foreign DNA linked to itself, vector DNA linked to itself and higher-order combinations of vector and foreign DNA) are also usually present. This complex mixture is sorted out in subsequent steps of the cloning process, after the DNA mixture is introduced into cells.

Introduction of Recombinant DNA into Host Organism

The DNA mixture, previously manipulated in vitro, is moved back into a living cell, referred to as the host organism. The methods used to get DNA into cells are varied, and the name applied to this step in the molecular cloning process will often depend upon the experimental method that is chosen (e.g. transformation, transduction, transfection, electroporation).

When microorganisms are able to take up and replicate DNA from their local environment, the process is termed transformation, and cells that are in a physiological state such that they can take up DNA are said to be competent. In mammalian cell culture, the analogous process of introducing DNA into cells is commonly termed transfection. Both transformation and transfection usually require preparation of the cells through a special growth regime and chemical treatment process that will vary with the specific species and cell types that are used.

Electroporation uses high voltage electrical pulses to translocate DNA across the cell membrane (and cell wall, if present). In contrast, transduction involves the packaging of DNA into virus-derived particles, and using these virus-like particles to introduce the encapsulated DNA into the cell through a process resembling viral infection. Although electroporation and transduction are highly specialised methods, they may be the most efficient methods to move DNA into cells.

Selection of Organisms Containing Vector Sequences

Whichever method is used, the introduction of recombinant DNA into the chosen host organism is usually a low efficiency process; that is, only a small fraction of the cells will actually take up DNA. Experimental scientists deal with this issue through a step of artificial genetic selection, in which cells that have not taken up DNA are selectively killed, and only those cells that can actively replicate DNA containing the selectable marker gene encoded by the vector are able to survive.

When bacterial cells are used as host organisms, the selectable marker is usually a gene that confers resistance to an antibiotic that would otherwise kill the cells, typically ampicillin. Cells harbouring the vector will survive when exposed to the antibiotic, while those that have failed to take up vector sequences will die. When mammalian cells (e.g. human or mouse cells) are used, a similar strategy is used, except that the marker gene confers resistance to the antibiotic Geneticin.

Screening for Clones with Desired DNA Inserts and Biological Properties

Modern bacterial cloning vectors (e.g. pUC19 and later derivatives including the pGEM vectors) use the blue-white screening system to distinguish colonies (clones) of transgenic cells from those that contain the parental vector (i.e. vector DNA with no recombinant sequence inserted). In these vectors, foreign DNA is inserted into a sequence that encodes an essential part of beta-galactosidase, an enzyme whose activity results in formation of a blue-coloured colony on the culture medium that is used for this work. Insertion of the foreign DNA into the beta-galactosidase coding sequence disables the function of the enzyme, so that colonies containing recombinant plasmids remain colourless (white). Therefore, experimentalists are easily able to identify and conduct further studies on transgenic bacterial clones, while ignoring those that do not contain recombinant DNA.

The total population of individual clones obtained in a molecular cloning experiment is often termed a DNA library. Libraries may be highly complex (as when cloning complete genomic DNA from an organism) or relatively simple (as when moving a previously-cloned DNA fragment into a different plasmid), but it is almost always necessary to examine a number of different clones to be sure that the desired DNA construct is obtained. This may be accomplished through a very wide range of experimental methods, including the use of nucleic acid hybridisations, antibody probes, polymerase chain reaction, restriction fragment analysis and/or DNA sequencing.

Applications of Molecular Cloning

Molecular cloning provides scientists with an essentially unlimited quantity of any individual DNA segments derived from any genome. This material can be used for a wide range of purposes, including those in both basic and applied biological science. A few of the more important applications are summarised here.

Genome Organisation and Gene Expression

Molecular cloning has led directly to the elucidation of the complete of sexual DNA sequence of the genomes of a very large number of species and to an exploration of genetic diversity within individual species, work that has been done mostly by determining the DNA sequence of large numbers of randomly cloned fragments of the genome, and assembling the overlapping sequences.

At the level of individual genes, molecular clones are used to generate probes that are used for examining how genes are expressed, and how that expression is related to other processes in biology, including the metabolic environment, extracellular signals, development, learning, senescence and cell death. Cloned genes can also provide tools to examine the biological function and importance of individual genes, by allowing investigators to inactivate the genes, or make more subtle mutations using regional mutagenesis or site-directed mutagenesis.

Production of Recombinant Proteins

Obtaining the molecular clone of a gene can lead to the development of organisms that produce the protein product of the cloned genes, termed a recombinant protein. In practice, it is frequently more difficult to develop an organism that produces an active form of the recombinant protein in desirable quantities than it is to clone the gene. This is because the molecular signals for gene expression are complex and variable, and because protein folding, stability and transport can be very challenging.

Many useful proteins are currently available as recombinant products. These include—(1) medically-useful proteins whose administration can correct a defective or poorly-expressed gene (e.g. recombinant factor VIII, a blood-clotting factor deficient in some forms of hemophilia, and recombinant insulin, used to treat some forms of diabetes), (2) proteins that can be administered to assist in a life threatening emergency (e.g. tissue plasminogen activator, used to treat strokes), and (3) recombinant subunit vaccines, in which a purified protein can be used to immunize patients against infectious diseases, without exposing them to the infectious agent itself (e.g. hepatitis B vaccine).

Transgenic Organisms

Once characterised and manipulated to provide signals for appropriate expression, cloned genes may be inserted into organisms, generating transgenic organisms, also termed genetically-modified organisms (GMOs). Although most GMOs are generated for purposes of basic biological research, a number of GMOs have been developed for commercial use, ranging from animals and plants that produce pharmaceuticals or other compounds (pharming), herbicide-resistant crop plants, and fluorescent tropical fish (GloFish) for home entertainment.

Gene Therapy

Gene therapy involves supplying a functional gene to cells lacking that function, with the aim of correcting a genetic disorder or acquired disease. Gene therapy can be broadly divided into two categories. The first is alteration of germ cells, that is, sperm or eggs, which results in a permanent genetic change for the whole organism and subsequent generations. This "germ line gene therapy" is considered by many to be unethical in human beings. The second type of gene therapy, "somatic cell gene therapy", is analogous to an organ transplant. In this case, one or more specific tissues are targeted by direct treatment or by removal of the tissue, addition of the therapeutic gene or genes in the laboratory, and return of the treated cells to the patient. Clinical trials of somatic cell gene therapy began in the late 1990s, mostly for the treatment of cancers and blood, liver, and lung disorders.

Despite a great deal of publicity and promises, the history of human gene therapy has been characterised by relatively limited success. The effect of introducing a gene into cells often promotes only partial and/or transient relief from the symptoms of the disease being treated. Some gene therapy trial patients have suffered adverse consequences of the treatment itself, including deaths. In some cases, the adverse effects result from disruption of essential genes within the patient's genome by insertional inactivation. In others, viral vectors used for gene therapy have been contaminated with infectious virus. Nevertheless, gene therapy is still held to be a promising future area of medicine, and is an area where there is a significant level of research and development activity.

Cloning of Viral Antigens

A virus is a small infectious agent that can replicate only inside the living cells of organisms. Viruses infect all types of organisms, from animals and plants to bacteria and archaea. Since Dmitri Ivanovsky's 1892 article describing a non-bacterial pathogen infecting tobacco plants, and the discovery of the tobacco mosaic virus by Martinus Beijerinck in 1898, about 5,000 viruses have been described in detail, although there are millions of different types. Viruses are found in almost every ecosystem on Earth and are the most abundant type of biological entity. The study of viruses is known as virology, a sub-speciality of microbiology.

Virus particles (known as *virions*) consist of two or three parts: the genetic material made from either DNA or RNA, long molecules that carry genetic information; a protein coat that protects these genes; and in some cases an envelope of lipids that surrounds the protein coat when they are outside a cell. The shapes of viruses range from simple helical and icosahedral forms to more complex structures. The average virus is about one one-hundredth the size of the average bacterium. Most viruses are too small to be seen directly with a light microscope.

The origins of viruses in the evolutionary history of life are unclear: some may have evolved from plasmids – pieces of DNA that can move between cells – while others may have evolved from bacteria. In evolution, viruses are an important means of horizontal gene transfer, which increases genetic diversity.

Viruses spread in many ways; viruses in plants are often transmitted from plant to plant by insects that feed on the sap of plants, such as aphids; viruses in animals can be carried by blood-sucking insects. These disease-bearing organisms are known as vectors. Influenza viruses are spread by coughing and sneezing. Norovirus and rotavirus, common causes of viral gastroenteritis, are transmitted by the faecal-oral route and are passed from person to person by contact, entering the body in food or water. HIV is one of several viruses transmitted through sexual contact and by exposure to infected blood. The range of host cells that a virus can infect is called its "host range". This can be narrow or, as when a virus is capable of infecting many species, broad.

Viral infections in animals provoke an immune response that usually eliminates the infecting virus. Immune responses can also be produced by vaccines, which confer an artificially acquired immunity to the specific viral infection. However, some viruses including those causing AIDS and viral hepatitis evade these immune responses and result in chronic infections. Antibiotics have no effect on viruses, but several antiviral drugs have been developed.

Etymology

The word is from the Latin *virus* referring to poison and other noxious substances, first used in English in 1392. *Virulent*, from Latin *virulentus* (poisonous), dates to 1400. A meaning of "agent that causes infectious disease" is first recorded in 1728, before the discovery of viruses by Dmitry Ivanovsky in 1892. The plural is *viruses*. The adjective *viral* dates to 1948. The term *virion* is also used to refer to a single infective viral particle.

History

Louis Pasteur was unable to find a causative agent for rabies and speculated about a pathogen too small to be detected using a microscope. In 1884, the French microbiologist Charles Chamberland invented a filter (known today as the Chamberland filter or Chamberland-Pasteur filter) with pores smaller than bacteria. Thus, he could pass a solution containing bacteria through the filter and completely remove them from the solution.

In 1892, the Russian biologist Dmitry Ivanovsky used this filter to study what is now known as the tobacco mosaic virus. His experiments showed that crushed leaf extracts from infected tobacco plants remain infectious after filtration. Ivanovsky suggested the infection might be caused by a toxin produced by bacteria, but did not pursue the idea. At the time it was thought that all infectious agents could be retained by filters and grown on a nutrient medium – this was part of the germ theory of disease. In 1898, the Dutch microbiologist Martinus Beijerinck repeated the experiments and became convinced that the filtered solution contained a new form of infectious agent. He observed that the agent multiplied only in cells that were dividing, but as his experiments did not show that it was made of particles, he called it a *contagium vivum fluidum* (soluble living germ) and reintroduced the word *virus*.

Beijerinck maintained that viruses were liquid in nature, a theory later discredited by Wendell Stanley, who proved they were particulate. In the same year Friedrich Loeffler and Frosch passed the first animal virus – agent of foot-and-mouth disease (aphthovirus) – through a similar filter.

In the early 20th century, the English bacteriologist Frederick Twort discovered a group of viruses that infect bacteria, now called bacteriophages (or commonly *phages*), and the French-Canadian microbiologist Félix d'Herelle described viruses that, when added to bacteria on agar, would produce areas of dead bacteria. He accurately diluted a suspension of these viruses and discovered that the highest dilutions (lowest virus concentrations), rather than killing all the bacteria, formed discrete areas of dead organisms.

Counting these areas and multiplying by the dilution factor allowed him to calculate the number of viruses in the original suspension. Phages were heralded as a potential treatment for diseases such as typhoid and cholera, but their promise was forgotten with the development of penicillin. The study of phages provided insights into the switching on and off of genes, and a useful mechanism for introducing foreign genes into bacteria.

By the end of the 19th century, viruses were defined in terms of their infectivity, their ability to be filtered, and their requirement for living hosts. Viruses had been grown only in plants and animals. In 1906, Ross Granville Harrison invented a method for growing tissue in lymph, and, in 1913, E. Steinhardt, C. Israeli, and R. A. Lambert used this method to grow vaccinia virus in fragments of guinea pig corneal tissue. In 1928, H. B. Maitland and M. C. Maitland grew vaccinia virus in suspensions of minced hens' kidneys. Their method was not widely adopted until the 1950s, when poliovirus was grown on a large scale for vaccine production. Another breakthrough came in 1931, when the American pathologist Ernest William Goodpasture grew influenza and several other viruses in fertilized chickens' eggs. In 1949, John F. Enders, Thomas Weller, and Frederick Robbins grew polio virus in cultured human embryo cells, the first virus to be grown without using solid animal tissue or eggs. This work enabled Jonas Salk to make an effective polio vaccine.

The first images of viruses were obtained upon the invention of electron microscopy in 1931 by the German engineers Ernst Ruska and Max Knoll. In 1935, American biochemist and virologist Wendell Meredith Stanley examined the tobacco mosaic virus and found it was mostly made of protein. A short time later, this virus was separated into protein and RNA parts. The tobacco mosaic virus was the first to be crystallised and its structure could therefore be elucidated in detail. The first X-ray diffraction pictures of the crystallised virus were obtained by Bernal and Fankuchen in 1941. On the basis of her pictures, Rosalind Franklin discovered the full DNA structure of the virus in 1955. In the same year, Heinz Fraenkel-Conrat and Robley Williams showed that purified tobacco mosaic virus RNA and its coat protein can assemble by themselves to form functional viruses, suggesting that this simple mechanism was probably the means through which viruses were created within their host cells.

The second half of the 20th century was the golden age of virus discovery and most of the 2,000 recognised species of animal, plant, and bacterial viruses were discovered during these years. In 1957, equine arterivirus and the cause of Bovine virus diarrhea (a pestivirus) were discovered. In 1963, the hepatitis B virus was discovered by Baruch Blumberg, and in 1965, Howard Temin described the first retrovirus. Reverse transcriptase, the key enzyme that retroviruses use to translate their RNA into DNA, was first described in 1970, independently by Howard Martin Temin and David Baltimore. In 1983 Luc Montagnier's team at the Pasteur Institute in France, first isolated the retrovirus now called HIV.

Origins

Viruses are found wherever there is life and have probably existed since living cells first evolved. The origin of viruses is unclear because they do not form fossils, so molecular techniques have been used to compare the DNA or RNA of viruses and are a useful means of investigating how they arose. There are three main hypotheses that try to explain the origins of viruses:

Regressive hypothesis : Viruses may have once been small cells that parasitised larger cells. Over time, genes not required by their parasitism were lost. The bacteria rickettsia and chlamydia are living cells that, like viruses, can reproduce only inside host cells. They lend support to this hypothesis, as their dependence on parasitism is likely to have caused the loss of genes that enabled them to survive outside a cell. This is also called the *degeneracy hypothesis*, or *reduction hypothesis*.

Cellular origin hypothesis: Some viruses may have evolved from bits of DNA or RNA that "escaped" from the genes of a larger organism. The escaped DNA could have come from plasmids (pieces of naked DNA that can move *between* cells) or transposons (molecules of DNA that replicate and move around to different positions *within* the genes of the cell). Once called "jumping genes", transposons are examples of mobile genetic elements and could be the origin of some viruses. They were discovered in maize by Barbara McClintock in 1950. This is sometimes called the *vagrancy hypothesis*, or the *escape hypothesis*.

Coevolution hypothesis: This is also called the *virus-first hypothesis* and proposes that viruses may have evolved from complex molecules of protein and nucleic acid at the same time as cells first appeared on earth and would have been dependent on cellular life for billions of years. Viroids are molecules of

RNA that are not classified as viruses because they lack a protein coat. However, they have characteristics that are common to several viruses and are often called subviral agents. Viroids are important pathogens of plants. They do not code for proteins but interact with the host cell and use the host machinery for their replication.

The hepatitis delta virus of humans has an RNA genome similar to viroids but has a protein coat derived from hepatitis B virus and cannot produce one of its own. It is, therefore, a defective virus and cannot replicate without the help of hepatitis B virus. In similar manner, the virophage 'sputnik' is dependent on mimivirus, which infects the protozoan *Acanthamoeba castellanii*. These viruses that are dependent on the presence of other virus species in the host cell are called *satellites* and may represent evolutionary intermediates of viroids and viruses.

In the past, there were problems with all of these hypotheses: the regressive hypothesis did not explain why even the smallest of cellular parasites do not resemble viruses in any way. The escape hypothesis did not explain the complex capsids and other structures on virus particles.

The virus-first hypothesis contravened the definition of viruses in that they require host cells. Viruses are now recognised as ancient and to have origins that predate the divergence of life into the three domains. This discovery has led modern virologists to reconsider and re-evaluate these three classical hypotheses.

The evidence for an ancestral world of RNA cells and computer analysis of viral and host DNA sequences are giving a better understanding of the evolutionary relationships between different viruses and may help identify the ancestors of modern viruses. To date, such analyses have not proved which of these hypotheses is correct. However, it seems unlikely that all currently known viruses have a common ancestor, and viruses have probably arisen numerous times in the past by one or more mechanisms.

Prions are infectious protein molecules that do not contain DNA or RNA. They cause an infection in sheep called scrapie and cattle bovine spongiform encephalopathy ("mad cow" disease). In humans they cause kuru and Creutzfeldt-Jakob disease. They are able to replicate because some proteins can exist in two different shapes and the prion changes the normal shape of a host protein into the prion shape.

This starts a chain reaction where each prion protein converts many host proteins into more prions, and these new prions then go on to convert even more protein into prions. Although they are fundamentally different from viruses and viroids, their discovery gives credence to the idea that viruses could have evolved from self-replicating molecules.

Microbiology

Life Properties

Opinions differ on whether viruses are a form of life, or organic structures that interact with living organisms. They have been described as "organisms at the edge of life", since they resemble organisms in that they possess genes and evolve by natural selection, and reproduce by creating multiple copies of themselves through self-assembly. Although they have genes, they do not have a cellular structure, which is often seen as the basic unit of life. Viruses do not have their own metabolism, and require a host cell to make new products. They therefore cannot naturally reproduce outside a host cell – although bacterial species such as rickettsia and chlamydia are considered living organisms despite the same limitation. Accepted forms of life use cell division to reproduce, whereas viruses spontaneously assemble within cells. They differ from autonomous growth of crystals as they inherit genetic mutations while being subject to natural selection. Virus self-assembly within host cells has implications for the study of the origin of life, as it lends further credence to the hypothesis that life could have started as self-assembling organic molecules.

Structure

Viruses display a wide diversity of shapes and sizes, called *morphologies*. Generally viruses are much smaller than bacteria. Most viruses that have been studied have a diameter between 20 and 300 nanometres. Some filoviruses have a total length of up to 1400 nm; their diameters are only about 80 nm. Most viruses cannot be seen with a light microscope so scanning and transmission electron microscopes are used to visualise virions. To increase the contrast between viruses and the background, electron-dense "stains" are used. These are solutions of salts of heavy metals, such as tungsten, that scatter the electrons from regions covered with the stain. When virions are coated with stain (positive staining), fine detail is obscured. Negative staining overcomes this problem by staining the background only.

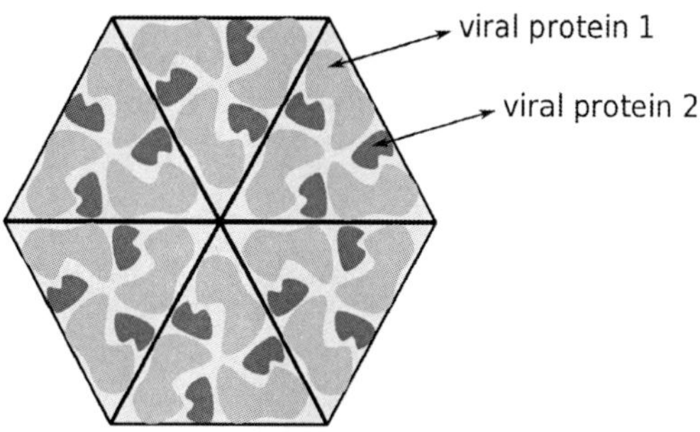

Figure: *Diagram of how a virus capsid can be constructed using multiple copies of just two protein molecules*

A complete virus particle, known as a virion, consists of nucleic acid surrounded by a protective coat of protein called a capsid. These are formed from identical protein subunits called capsomeres. Viruses can have a lipid "envelope" derived from the host cell membrane. The capsid is made from proteins encoded by the viral genome and its shape serves as the basis for morphological distinction. Virally coded protein subunits will self-assemble to form a capsid, generally requiring the presence of the virus genome. Complex viruses code for proteins that assist in the construction of their capsid. Proteins associated with nucleic acid are known as nucleoproteins, and the association of viral capsid proteins with viral nucleic acid is called a nucleocapsid. The capsid and entire virus structure can be mechanically (physically) probed through atomic force microscopy. In general, there are four main morphological virus types:

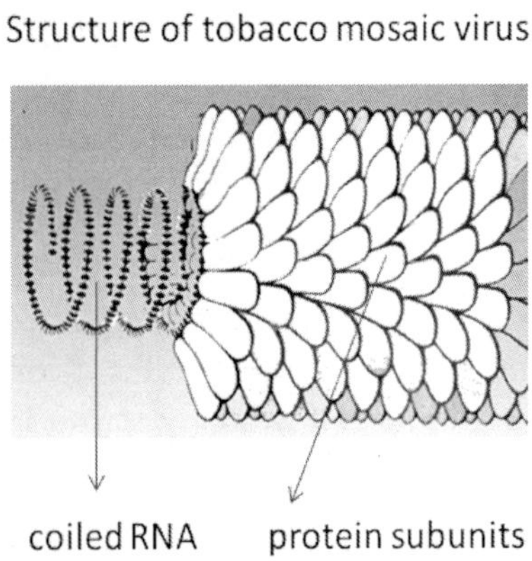

Figure: *RNA coiled in a helix of repeating protein subunits*

Helical: These viruses are composed of a single type of capsomer stacked around a central axis to form a helical structure, which may have a central cavity, or hollow tube. This arrangement results in rod-shaped or filamentous virions: These can be short and highly rigid, or long and very flexible.

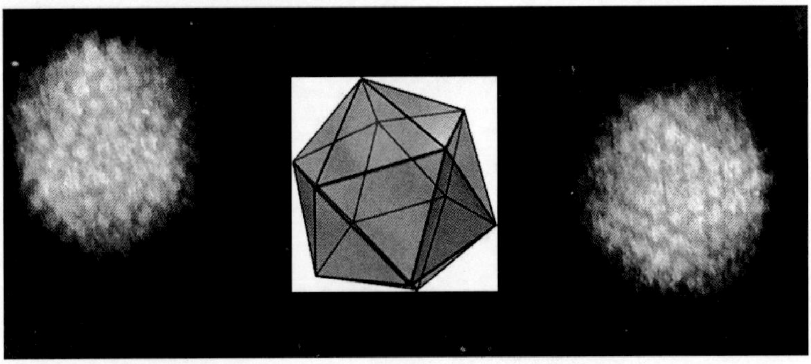

Figure: *Electron micrograph of icosahedral adenovirus*

The genetic material, in general, single-stranded RNA, but ssDNA in some cases, is bound into the protein helix by interactions between the negatively charged nucleic acid and positive charges on the protein. Overall, the length of a helical capsid is related to the length of the nucleic acid contained within it and the diameter is dependent on the size and arrangement of capsomers. The well-studied tobacco mosaic virus is an example of a helical virus.

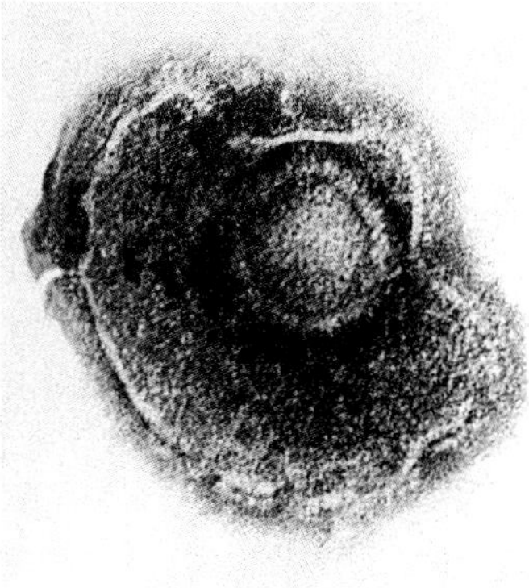

Figure: *Herpes viruses have a lipid envelope*

Icosahedral: Most animal viruses are icosahedral or near-spherical with icosahedral symmetry. A regular icosahedron is the optimum way of forming a closed shell from identical subunits. The minimum number of identical capsomers required is twelve, each composed of five identical subunits.

Many viruses, such as rotavirus, have more than twelve capsomers and appear spherical but they retain this symmetry. Capsomers at the apices are surrounded by five other capsomers and are called pentons. Capsomers on the triangular faces are surrounded by six others and are called hexons. Hexons are essentially flat and pentons, which form the 12 vertices, are curved. The same protein may act as the subunit of both the pentamers and hexamers or they may be composed of different proteins.

Although the isosahedral structure is extremely common among viruses because of size differences slight variations exist between virions. Given an asymmetric subunit on a triangular face of a regular isosahedron, with three subunits per face 60 such subunits can be placed in an equivalent manner. Most virions, because of their size, have more than 60 subunits. These variations have been classified on the basis of the quasi-equivalence principle proposed by Caspar and Klug.

An isosahedral structure can be regarded as being constructed from 12 pentamers. The number of pentamers is fixed but the number of hexamers can vary. These shells can be constructed from pentamers and hexamers by minimising the number T (triangulation number) of nonequivalent locations that subunits occupy, with the T-number adopting the particular integer values 1, 3, 4, 7, 12, 13,...(T = $h^2 + k^2 + hk$, with h, k equal to nonnegative integers). These shells always contain 12 pentamers plus 10 (T-1) hexamers. Although this classification can be applied to the majority of known viruses exceptions are known including the retroviruses where point mutations disrupt the symmetry.

Prolate: This is an isosahedron elongated along the fivefold axis and is a common arrangement of the heads of bacteriophages. Such a structure is composed of a cylinder with a cap at either end. The cylinder is composed of 10 triangles. The Q number, which can be can be any positive integer, specifies the number of triangles, composed of asymmetric subunits, that make up the 10 triangles of the cylinder. The caps are classified by the T number.

Envelope: Some species of virus envelop themselves in a modified form of one of the cell membranes, either the outer membrane surrounding an infected host cell or internal membranes such as nuclear membrane or endoplasmic reticulum, thus gaining an outer lipid bilayer known as a viral envelope. This membrane is studded with proteins coded for by the viral genome and host genome; the lipid membrane itself and any carbohydrates present originate entirely from the host. The influenza virus and HIV use this strategy. Most enveloped viruses are dependent on the envelope for their infectivity.

Complex: These viruses possess a capsid that is neither purely helical nor purely icosahedral, and that may possess extra structures such as protein tails or a complex outer wall. Some bacteriophages, such as Enterobacteria phage T4, have a complex structure consisting of an icosahedral head bound to a helical tail, which may have a hexagonal base plate with protruding protein tail fibres. This tail structure acts like a molecular syringe, attaching to the bacterial host and then injecting the viral genome into the cell.

The poxviruses are large, complex viruses that have an unusual morphology. The viral genome is associated with proteins within a central disk structure known as a nucleoid. The nucleoid is surrounded by a membrane and two lateral bodies of unknown function. The virus has an outer envelope with a thick layer of protein studded over its surface.

The whole virion is slightly pleiomorphic, ranging from ovoid to brick shape. Mimivirus is the largest characterised virus, with a capsid diameter of 400 nm. Protein filaments measuring 100 nm project from the surface. The capsid appears hexagonal under an electron microscope, therefore the capsid is probably icosahedral. In 2011, researchers discovered a larger virus on ocean floor of the coast of Las Cruces, Chile. Provisionally named *Megavirus chilensis*, it can be seen with a basic light microscope. Some viruses that infect Archaea have complex structures that are unrelated to any other form of virus, with a wide variety of unusual shapes, ranging from spindle-shaped structures, to viruses that resemble hooked rods, teardrops or even bottles. Other archaeal viruses resemble the tailed bacteriophages, and can have multiple tail structures.

Genome

An enormous variety of genomic structures can be seen among viral species; as a group they contain more structural genomic diversity than plants, animals, archaea, or bacteria. There are millions of different types of viruses, although only about 5,000 of them have been described in detail. A virus has either DNA or RNA genes and is called a DNA virus or a RNA virus respectively. The vast majority of viruses have RNA genomes. Plant viruses tend to have single-stranded RNA genomes and bacteriophages tend to have double-stranded DNA genomes.

Viral genomes are circular, as in the polyomaviruses, or linear, as in the adenoviruses. The type of nucleic acid is irrelevant to the shape of the genome. Among RNA viruses and certain DNA viruses, the genome is often divided up into separate parts, in which case it is called *segmented*. For RNA viruses, each segment often codes for only one protein and they are usually found together in one capsid. However, all segments are not required to be in the same virion for the virus to be infectious, as demonstrated by brome mosaic virus and several other plant viruses.

A viral genome, irrespective of nucleic acid type, is almost always either single-stranded or double-stranded. Single-stranded genomes consist of an unpaired nucleic acid, analogous to one-half of a ladder split down the middle. Double-stranded genomes consist of two complementary paired nucleic acids, analogous to a ladder. The virus particles of some virus families, such as those belonging to the *Hepadnaviridae*, contain a genome that is partially double-stranded and partially single-stranded. For most viruses with RNA genomes and some with single-stranded DNA genomes, the single strands are said to be either positive-sense (called the plus-strand) or negative-sense (called the minus-strand), depending on whether or not they are complementary to the viral messenger RNA (mRNA). Positive-sense viral RNA is in the same sense as viral mRNA and thus at least a part of it can be immediately translated by the host cell. Negative-sense viral RNA is complementary to mRNA and thus must be converted to positive-sense RNA by an RNA-dependent RNA polymerase before translation.

DNA nomenclature for viruses with single-sense genomic ssDNA is similar to RNA nomenclature, in that the *coding strand* for the viral mRNA is complementary to it ("), and the *non-coding strand* is a copy of it (+). However, several types of ssDNA and ssRNA viruses have genomes that are ambisense in that transcription can occur off both strands in a double-stranded replicative intermediate. Examples include geminiviruses, which are ssDNA plant viruses and arenaviruses, which are ssRNA viruses of animals.

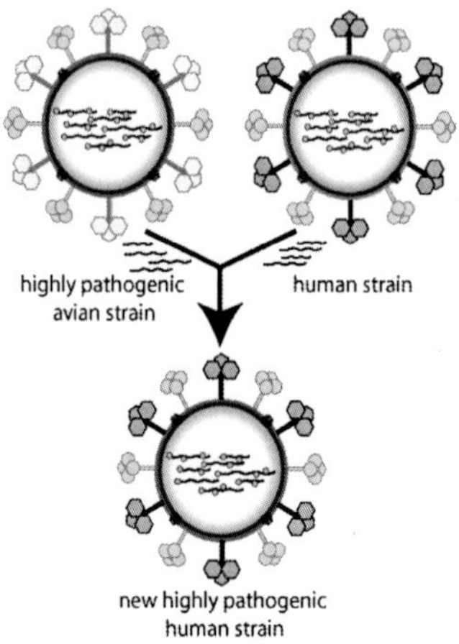

Figure: *How antigenic shift, or reassortment, can result in novel and highly pathogenic strains of human influenza*

Genome size varies greatly between species. The smallest viral genomes – the ssDNA circoviruses, family *Circoviridae* – code for only two proteins and have a genome size of only 2 kilobases; the largest – mimiviruses – have genome sizes of over 1.2 megabases and code for over one thousand proteins. RNA viruses generally have smaller genome sizes than DNA viruses because of a higher error-rate when replicating, and have a maximum upper size limit. Beyond this limit, errors in the genome when replicating render the virus useless or uncompetitive. To compensate for this, RNA viruses often have segmented genomes – the genome is split into smaller molecules – thus reducing the chance that an error in a single-component genome will incapacitate the entire genome. In contrast, DNA viruses generally have larger genomes because of the high fidelity of their replication enzymes. Single-strand DNA viruses are an exception to this rule, however, as mutation rates for these genomes can approach the extreme of the ssRNA virus case.

Viruses undergo genetic change by several mechanisms. These include a process called genetic drift where individual bases in the DNA or RNA mutate to other bases. Most of these point mutations are "silent" – they do not change the protein that the gene encodes – but others can confer evolutionary advantages such as resistance to antiviral drugs. Antigenic shift occurs when there is a major change in the genome of the virus. This can be a result of recombination or reassortment. When this happens with influenza viruses, pandemics might result. RNA viruses often exist as quasispecies or swarms of viruses of the same species but with slightly different genome nucleoside sequences. Such quasispecies are a prime target for natural selection.

Segmented genomes confer evolutionary advantages; different strains of a virus with a segmented genome can shuffle and combine genes and produce progeny viruses or (offspring) that have unique characteristics. This is called reassortment or *viral sex.* Genetic recombination is the process by which a strand of DNA is broken and then joined to the end of a different DNA molecule. This can occur when viruses infect cells simultaneously and studies of viral evolution have shown that recombination has been rampant in the species studied. Recombination is common to both RNA and DNA viruses.

Replication Cycle

Viral populations do not grow through cell division, because they are acellular. Instead, they use the machinery and metabolism of a host cell to produce multiple copies of themselves, and they *assemble* in the cell.

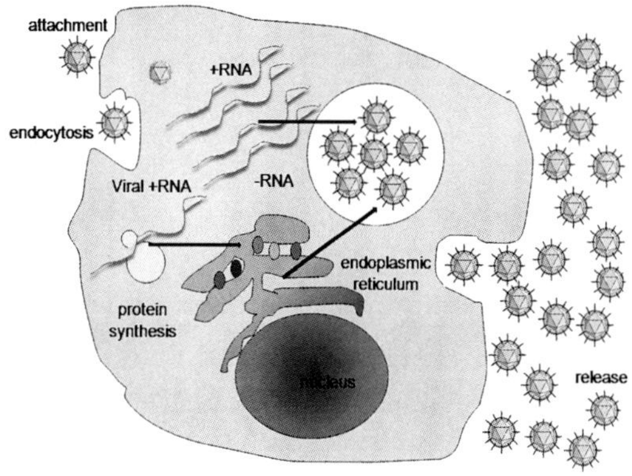

Figure: *A typical virus replication cycle*

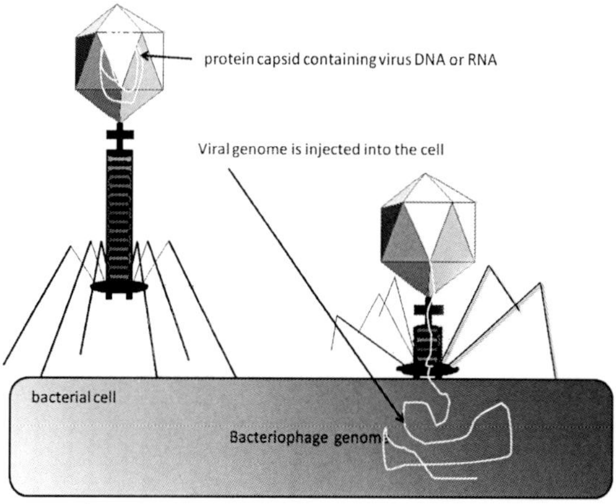

Figure: *Some bacteriophages inject their genomes into bacterial cells*

The life cycle of viruses differs greatly between species but there are six *basic* stages in the life cycle of viruses:

- *Attachment* is a specific binding between viral capsid proteins and specific receptors on the host cellular surface. This specificity determines the host range of a virus. For example, HIV infects a limited range of human leucocytes. This is because its surface protein, gp120, specifically interacts with the CD4 molecule – a chemokine receptor – which is most commonly found on the surface of CD4+ T-Cells. This mechanism has evolved to favour those viruses that infect only cells in which they are capable of replication. Attachment to the receptor can induce the viral envelope protein to undergo changes that results in the fusion of viral and cellular membranes, or changes of non-enveloped virus surface proteins that allow the virus to enter.
- *Penetration* follows attachment: Virions enter the host cell through receptor-mediated endocytosis or membrane fusion. This is often called viral entry. The infection of plant and, it is presumed, fungal cells is different from that of animal cells. Plants have a rigid cell wall made of cellulose, and fungi one of chitin, so most viruses can get inside these cells only after trauma to the cell wall. However, nearly all plant viruses (such as tobacco mosaic virus) can also move directly from cell to cell, in the form of single-stranded nucleoprotein complexes, through pores called plasmodesmata. This process requires movement proteins, which are virus-encoded proteins probably originally derived from plant proteins, which interact with the plasmodesmatal transport machinery Bacteria, like plants, have strong cell walls that a virus must breach to infect the cell. However, given that bacterial cell walls are much less thick than plant cell walls due to their much smaller size, some viruses have evolved mechanisms that inject their genome into the bacterial cell across the cell wall, while the viral capsid remains outside.
- *Uncoating* is a process in which the viral capsid is removed: This may be by degradation by viral enzymes or host enzymes or by simple dissociation; the end-result is the releasing of the viral genomic nucleic acid.
- *Replication* of viruses primarily involves multiplication of the genome. Replication involves synthesis of viral messenger RNA (mRNA) from "early" genes (with exceptions for positive sense RNA viruses), viral protein synthesis, possible assembly of viral proteins, then viral genome replication mediated by early or regulatory protein expression. This may be followed, for complex viruses with larger genomes, by one or more further rounds of mRNA synthesis: "late" gene expression is, in general, of structural or virion proteins.

- Following the structure-mediated self-*assembly* of the virus particles, some modification of the proteins often occurs. In viruses such as HIV, this modification (sometimes called maturation) occurs *after* the virus has been released from the host cell.
- Viruses can be *released* from the host cell by lysis, a process that kills the cell by bursting its membrane and cell wall if present: This is a feature of many bacterial and some animal viruses. Some viruses undergo a lysogenic cycle where the viral genome is incorporated by genetic recombination into a specific place in the host's chromosome. The viral genome is then known as a "provirus" or, in the case of bacteriophages a "prophage". Whenever the host divides, the viral genome is also replicated. The viral genome is mostly silent within the host; however, at some point, the provirus or prophage may give rise to active virus, which may lyse the host cells. Enveloped viruses (e.g., HIV) typically are released from the host cell by budding. During this process the virus acquires its envelope, which is a modified piece of the host's plasma or other, internal membrane.

The genetic material within virus particles, and the method by which the material is replicated, varies considerably between different types of viruses.

DNA viruses: The genome replication of most DNA viruses takes place in the cell's nucleus. If the cell has the appropriate receptor on its surface, these viruses enter the cell sometimes by direct fusion with the cell membrane (e.g. herpesviruses) or – more usually – by receptor-mediated endocytosis. Most DNA viruses are entirely dependent on the host cell's DNA and RNA synthesising machinery, and RNA processing machinery; however, viruses with larger genomes may encode much of this machinery themselves. In eukaryotes the viral genome must cross the cell's nuclear membrane to access this machinery, while in bacteria it need only enter the cell.

RNA viruses: Replication usually takes place in the cytoplasm. RNA viruses can be placed into four different groups depending on their modes of replication. The polarity (whether or not it can be used directly by ribosomes to make proteins) of single-stranded RNA viruses largely determines the replicative mechanism; the other major criterion is whether the genetic material is single-stranded or double-stranded. All RNA viruses use their own RNA replicase enzymes to create copies of their genomes.

Reverse transcribing viruses: These have ssRNA (*Retroviridae*, *Metaviridae*, *Pseudoviridae*) or dsDNA (*Caulimoviridae*, and *Hepadnaviridae*) in their particles. Reverse transcribing viruses with RNA genomes (retroviruses), use a DNA intermediate to replicate, whereas those with DNA genomes (pararetroviruses) use an RNA intermediate during genome replication. Both types use a reverse transcriptase, or RNA-dependent DNA polymerase enzyme, to carry out the nucleic acid conversion. Retroviruses integrate the DNA produced by reverse transcription into the host genome as a provirus as a part of the replication process; pararetroviruses do not, although integrated genome copies of especially plant pararetroviruses can give rise to infectious virus. They are susceptible to antiviral drugs that inhibit the reverse transcriptase enzyme, e.g. zidovudine and lamivudine. An example of the first type is HIV, which is a retrovirus. Examples of the second type are the *Hepadnaviridae*, which includes Hepatitis B virus.

Effects on the Host Cell

The range of structural and biochemical effects that viruses have on the host cell is extensive. These are called *cytopathic effects*. Most virus infections eventually result in the death of the host cell. The causes of death include cell lysis, alterations to the cell's surface membrane and apoptosis. Often cell death is caused by cessation of its normal activities because of suppression by virus-specific proteins, not all of which are components of the virus particle.

Some viruses cause no apparent changes to the infected cell. Cells in which the virus is latent and inactive show few signs of infection and often function normally. This causes persistent infections and the virus is often dormant for many months or years. This is often the case with herpes viruses. Some viruses, such as Epstein-Barr virus, can cause cells to proliferate without causing malignancy, while others, such as papillomaviruses, are established causes of cancer.

Host Range

Viruses are by far the most abundant parasites on earth, and they have been found to infect all types of cellular life including animals, plants, bacteria and fungi. However, different types of viruses can infect only a limited range of hosts and many are species-specific. Some, such as smallpox virus for example, can infect only one species – in this case humans, and are said to have a narrow host range. Other viruses, such as rabies virus, can infect different species of mammals and are said to have a broad range. The viruses that infect plants are harmless to animals, and most viruses that infect other animals are harmless to humans. The host range of some bacteriophages is limited to a single strain of bacteria and they can be used to trace the source of outbreaks of infections by a method called phage typing.

Classification

Classification seeks to describe the diversity of viruses by naming and grouping them on the basis of similarities. In 1962, André Lwoff, Robert Horne, and Paul Tournier were the first to develop a means of virus classification, based on the Linnaean hierarchical system. This system bases classification on phylum, class, order, family, genus, and species. Viruses were grouped according to their shared properties (not those of their hosts) and the type of nucleic acid forming their genomes. Later the International Committee on Taxonomy of Viruses was formed. However, viruses are not classified on the basis of phylum or class, as their small genome size and high rate of mutation makes it difficult to determine their ancestry beyond Order. As such, the Baltimore Classification is used to supplement the more traditional hierarchy.

ICTV Classification

The International Committee on Taxonomy of Viruses (ICTV) developed the current classification system and wrote guidelines that put a greater weight on certain virus properties to maintain family uniformity. A unified taxonomy (a universal system for classifying viruses) has been established.

The 7th lCTV Report formalised for the first time the concept of the virus species as the lowest taxon (group) in a branching hierarchy of viral taxa. However, at present only a small part of the total diversity of viruses has been studied, with analyses of samples from humans finding that about 20% of the virus sequences recovered have not been seen before, and samples from the environment, such as from seawater and ocean sediments, finding that the large majority of sequences are completely novel.

The general taxonomic structure is as follows:

- Order (-virales)
- Family (-viridae)
- Subfamily (-virinae)
- Genus (-*virus*)
- Species (-*virus*).

In the current (2010) ICTV taxonomy, six orders have been established, the Caudovirales, Herpesvirales, Mononegavirales, Nidovirales, Picornavirales and Tymovirales. A seventh order Ligamenvirales has also been proposed. The committee does not formally distinguish between subspecies, strains, and isolates. In total there are 6 orders, 87 families, 19 subfamilies, 348 genera, 2,885 species and about 3,000 types yet unclassified.

Baltimore Classification

The Nobel Prize-winning biologist David Baltimore devised the Baltimore classification system. The ICTV classification system is used in conjunction with the Baltimore classification system in modern virus classification. The Baltimore classification of viruses is based on the mechanism of mRNA production. Viruses must generate mRNAs from their genomes to produce proteins and replicate themselves, but different mechanisms are used to achieve this in each virus family.

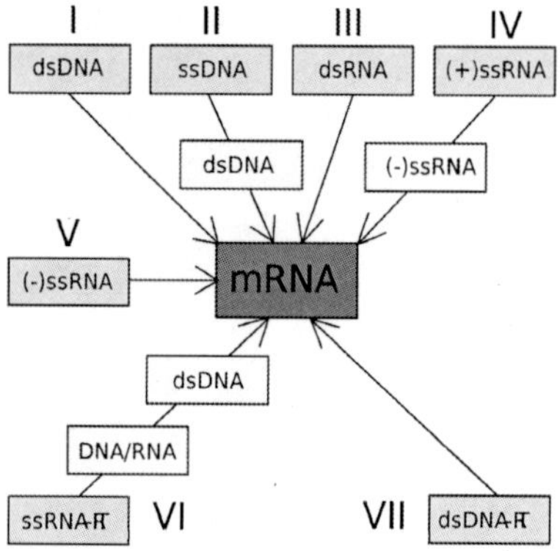

Figure: *The Baltimore Classification of viruses is based on the method of viral mRNA synthesis.*

Viral genomes may be single-stranded (ss) or double-stranded (ds), RNA or DNA, and may or may not use reverse transcriptase (RT). Additionally, ssRNA viruses may be either sense (+) or antisense ("). This classification places viruses into seven groups:

- I: dsDNA viruses (e.g. Adenoviruses, Herpesviruses, Poxviruses)
- II: ssDNA viruses (+)sense DNA (e.g. Parvoviruses)
- III: dsRNA viruses (e.g. Reoviruses)
- IV: (+)ssRNA viruses (+)sense RNA (e.g. Picornaviruses, Togaviruses)
- V: (")ssRNA viruses (—)sense RNA (e.g. Orthomyxoviruses, Rhabdoviruses)
- VI: ssRNA-RT viruses (+)sense RNA with DNA intermediate in life-cycle (e.g. Retroviruses)
- VII: dsDNA-RT viruses (e.g. Hepadnaviruses).

Role in Human Disease

As an example of viral classification, the chicken pox virus, varicella zoster (VZV), belongs to the order Herpesvirales, family *Herpesviridae*, subfamily *Alphaherpesvirinae*, and genus *Varicellovirus*. VZV is in Group I of the Baltimore Classification because it is a dsDNA virus that does not use reverse transcriptase.

Examples of common human diseases caused by viruses include the common cold, influenza, chickenpox and cold sores. Many serious diseases such as ebola, AIDS, avian influenza and SARS are caused by viruses. The relative ability of viruses to cause disease is described in terms of virulence. Other diseases are under investigation as to whether they too have a virus as the causative agent, such as the possible connection between human herpes virus six (HHV6) and neurological diseases such as multiple sclerosis and chronic fatigue syndrome. There is controversy over whether the borna virus, previously thought to cause neurological diseases in horses, could be responsible for psychiatric illnesses in humans.

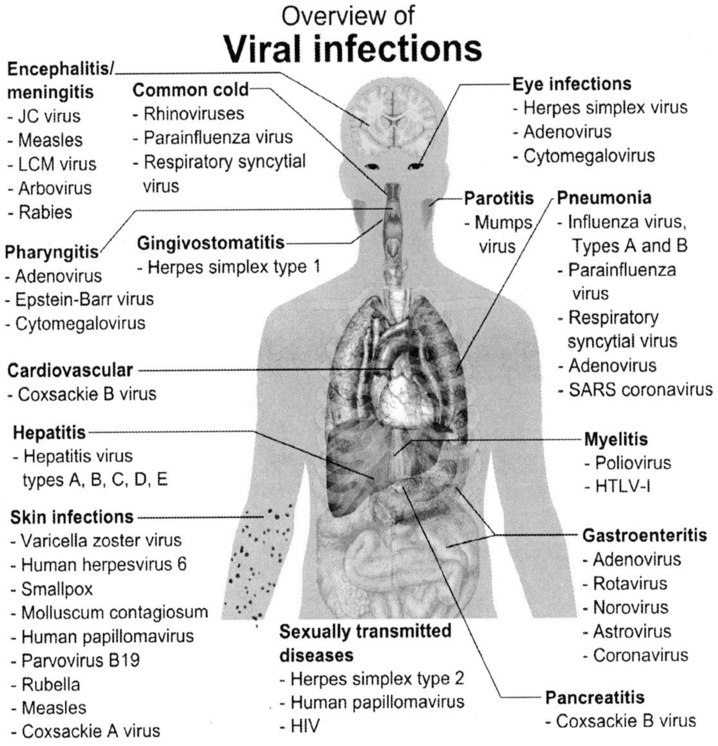

Figure: *Overview of the main types of viral infection and the most notable species involved.*

Viruses have different mechanisms by which they produce disease in an organism, which largely depends on the viral species. Mechanisms at the cellular level primarily include cell lysis, the breaking open and subsequent death of the cell. In multicellular organisms, if enough cells die, the whole organism will start to suffer the effects.

Although viruses cause disruption of healthy homeostasis, resulting in disease, they may exist relatively harmlessly within an organism. An example would include the ability of the herpes simplex virus, which causes cold sores, to remain in a dormant state within the human body.

This is called latency and is a characteristic of the herpes viruses including Epstein-Barr virus, which causes glandular fever, and varicella zoster virus, which causes chickenpox and shingles. Most people have been infected with at least one of these types of herpes virus. However, these latent viruses might sometimes be beneficial, as the presence of the virus can increase immunity against bacterial pathogens, such as *Yersinia pestis*. Some viruses can cause lifelong or chronic infections, where the viruses continue to replicate in the body despite the host's defence mechanisms. This is common in hepatitis B virus and hepatitis C virus infections. People chronically infected are known as carriers, as they serve as reservoirs of infectious virus. In populations with a high proportion of carriers, the disease is said to be endemic.

Epidemiology

Viral epidemiology is the branch of medical science that deals with the transmission and control of virus infections in humans. Transmission of viruses can be vertical, that is from mother to child, or horizontal, which means from person to person. Examples of vertical transmission include hepatitis B virus and HIV where the baby is born already infected with the virus. Another, more rare, example is the varicella zoster virus, which, although causing relatively mild infections in humans, can be fatal to the foetus and newborn baby.

Horizontal transmission is the most common mechanism of spread of viruses in populations. Transmission can occur when: body fluids are exchanged during sexual activity, e.g., HIV; blood is exchanged by contaminated transfusion or needle sharing, e.g., hepatitis C; a child is born to an infected mother, e.g., hepatitis B; exchange of saliva by mouth, e.g., Epstein-Barr virus; contaminated food or water is ingested, e.g., norovirus; aerosols containing virions are inhaled, e.g., influenza virus; and insect vectors such as mosquitoes penetrate the skin of a host, e.g., dengue. The rate or speed of transmission of viral infections depends on factors that include population density, the number of susceptible individuals, (i.e., those not immune), the quality of healthcare and the weather.

Epidemiology is used to break the chain of infection in populations during outbreaks of viral diseases. Control measures are used that are based on knowledge of how the virus is transmitted. It is important to find the source, or sources, of the outbreak and to identify the virus. Once the virus has been identified, the chain of transmission can sometimes be broken by vaccines. When vaccines are not available sanitation and disinfection can be effective.

Often infected people are isolated from the rest of the community and those that have been exposed to the virus placed in quarantine. To control the outbreak of foot and mouth disease in cattle in Britain in 2001, thousands of cattle were slaughtered.

Most viral infections of humans and other animals have incubation periods during which the infection causes no signs or symptoms. Incubation periods for viral diseases range from a few days to weeks but are known for most infections. Somewhat overlapping, but mainly following the incubation period, there is a period of communicability; a time when an infected individual or animal is contagious and can infect another person or animal. This too is known for many viral infections and knowledge the length of both periods is important in the control of outbreaks. When outbreaks cause an unusually high proportion of cases in a population, community or region they are called epidemics. If outbreaks spread worldwide they are called pandemics.

Epidemics and Pandemics

Native American populations were devastated by contagious diseases, in particular, smallpox, brought to the Americas by European colonists. It is unclear how many Native Americans were killed by foreign diseases after the arrival of Columbus in the Americas, but the numbers have been estimated to be close to 70% of the indigenous population. The damage done by this disease significantly aided European attempts to displace and conquer the native population.

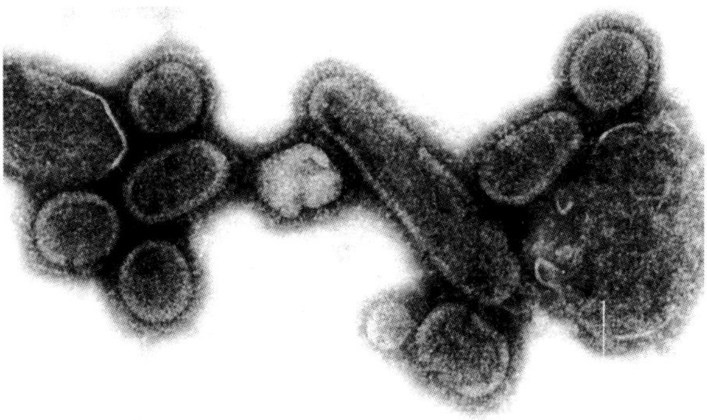

Figure: *Transmission electron microscope image of a recreated 1918 influenza virus*

A pandemic is a worldwide epidemic. The 1918 flu pandemic, commonly referred to as the Spanish flu, was a category 5 influenza pandemic caused by an unusually severe and deadly influenza A virus. The victims were often healthy young adults, in contrast to most influenza outbreaks, which predominantly affect juvenile, elderly, or otherwise-weakened patients.

The Spanish flu pandemic lasted from 1918 to 1919. Older estimates say it killed 40–50 million people, while more recent research suggests that it may have killed as many as 100 million people, or 5% of the world's population in 1918.

Most researchers believe that HIV originated in sub-Saharan Africa during the 20th century; it is now a pandemic, with an estimated 38.6 million people now living with the disease worldwide. The Joint United Nations Programme on HIV/AIDS (UNAIDS) and the World Health Organisation (WHO) estimate that AIDS has killed more than 25 million people since it was first recognised on June 5, 1981, making it one of the most destructive epidemics in recorded history. In 2007 there were 2.7 million new HIV infections and 2 million HIV-related deaths.

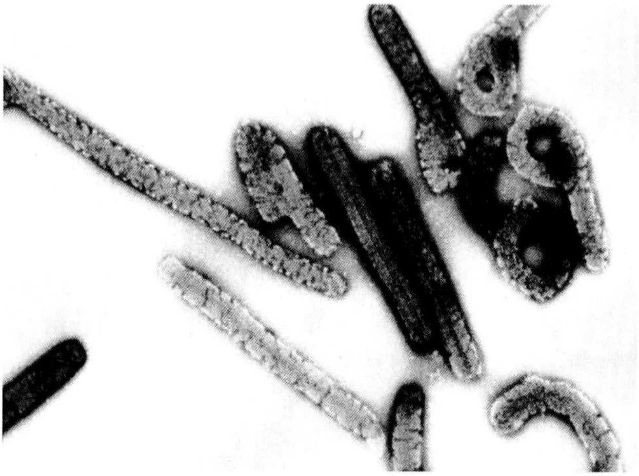

Figure: *Marburg virus*

Several highly lethal viral pathogens are members of the *Filoviridae*. Filoviruses are filament-like viruses that cause viral hemorrhagic fever, and include the ebola and marburg viruses. The Marburg virus attracted widespread press attention in April 2005 for an outbreak in Angola. Beginning in October 2004 and continuing into 2005, the outbreak was the world's worst epidemic of any kind of viral hemorrhagic fever.

Cancer

Viruses are an established cause of cancer in humans and other species. Viral cancers occur only in a minority of infected persons (or animals). Cancer viruses come from a range of virus families, including both RNA and DNA viruses, and so there is no single type of "oncovirus" (an obsolete term originally used for acutely transforming retroviruses). The development of cancer is determined by a variety of factors such as host immunity and mutations in the host. Viruses accepted to cause human cancers include some genotypes of human papillomavirus, hepatitis B virus, hepatitis C virus, Epstein-Barr virus, Kaposi's sarcoma-associated herpesvirus and human T-lymphotropic virus. The most recently discovered human cancer virus is a polyomavirus (Merkel cell polyomavirus) that causes most cases of a rare form of skin cancer called Merkel cell carcinoma. Hepatitis viruses can develop into a chronic viral infection that leads to liver cancer. Infection by human T-lymphotropic virus can lead to tropical spastic paraparesis and adult T-cell leukemia. Human papillomaviruses are an established cause of cancers of cervix, skin, anus, and penis. Within the *Herpesviridae*, Kaposi's sarcoma-associated herpesvirus causes Kaposi's sarcoma and body cavity lymphoma, and Epstein–Barr virus causes Burkitt's lymphoma, Hodgkin's lymphoma, B lymphoproliferative disorder, and nasopharyngeal carcinoma. Merkel cell polyomavirus closely related to SV40 and mouse polyomaviruses that have been used as animal models for cancer viruses for over 50 years.

Host Defence Mechanisms

The body's first line of defence against viruses is the innate immune system. This comprises cells and other mechanisms that defend the host from infection in a non-specific manner. This means that the cells of the innate system recognise, and respond to, pathogens in a generic way, but, unlike the adaptive immune system, it does not confer long-lasting or protective immunity to the host.

RNA interference is an important innate defence against viruses. Many viruses have a replication strategy that involves double-stranded RNA (dsRNA). When such a virus infects a cell, it releases its RNA molecule or molecules, which immediately bind to a protein complex called dicer that cuts the RNA into smaller pieces. A biochemical pathway called the RISC complex is activated, which degrades the viral mRNA and the cell survives the infection. Rotaviruses avoid this mechanism by not uncoating fully inside the cell and by releasing newly produced mRNA through pores in the particle's inner capsid. The genomic dsRNA remains protected inside the core of the virion.

When the adaptive immune system of a vertebrate encounters a virus, it produces specific antibodies that bind to the virus and render it non-infectious. This is called humoral immunity. Two types of antibodies are important. The first, called IgM, is highly effective at neutralizing viruses but is produced by the cells of the immune system only for a few weeks. The second, called IgG, is produced indefinitely. The presence of IgM in the blood of the host is used to test for acute infection, whereas IgG indicates an infection sometime in the past. IgG antibody is measured when tests for immunity are carried out. A second defence of vertebrates against viruses is called cell-mediated immunity and involves immune cells known as T cells. The body's cells constantly display short fragments of their proteins on the cell's surface, and, if a T cell recognises a suspicious viral fragment there, the host cell is destroyed by *killer T* cells and the virus-specific T-cells proliferate.

Cells such as the macrophage are specialists at this antigen presentation. The production of interferon is an important host defence mechanism. This is a hormone produced by the body when viruses are present. Its role in immunity is complex; it eventually stops the viruses from reproducing by killing the infected cell and its close neighbours.

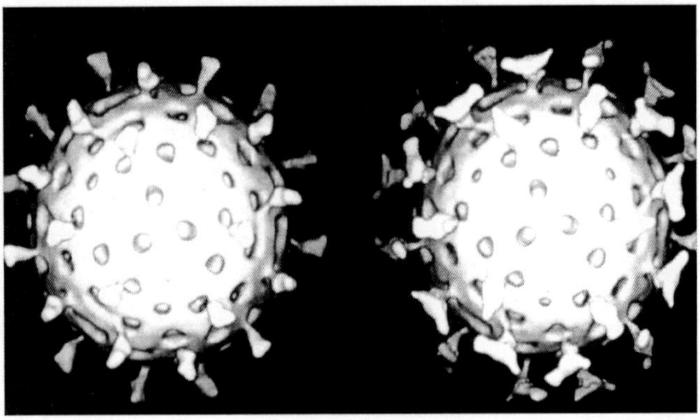

Figure: Two rotaviruses: the one on the right is coated with antibodies that stop its attaching to cells and infecting them

Not all virus infections produce a protective immune response in this way. HIV evades the immune system by constantly changing the amino acid sequence of the proteins on the surface of the virion. These persistent viruses evade immune control by sequestration, blockade of antigen presentation, cytokine resistance, evasion of natural killer cell activities, escape from apoptosis, and antigenic shift. Other viruses, called *neurotropic viruses*, are disseminated by neural spread where the immune system may be unable to reach them.

Prevention and Treatment

Because viruses use vital metabolic pathways within host cells to replicate, they are difficult to eliminate without using drugs that cause toxic effects to host cells in general. The most effective medical approaches to viral diseases are vaccinations to provide immunity to infection, and antiviral drugs that selectively interfere with viral replication.

Vaccines

Vaccination is a cheap and effective way of preventing infections by viruses. Vaccines were used to prevent viral infections long before the discovery of the actual viruses. Their use has resulted in a dramatic decline in morbidity (illness) and mortality (death) associated with viral infections such as polio, measles, mumps and rubella.

Antiviral Drugs

Smallpox infections have been eradicated. Vaccines are available to prevent over thirteen viral infections of humans, and more are used to prevent viral infections of animals. Vaccines can consist of live-attenuated or killed viruses, or viral proteins (antigens). Live vaccines contain weakened forms of the virus, which do not cause the disease but, nonetheless, confer immunity. Such viruses are called attenuated. Live vaccines can be dangerous when given to people with a weak immunity, (who are described as immunocompromised), because in these people, the weakened virus can cause the original disease. Biotechnology and genetic engineering techniques are used to produce subunit vaccines. These vaccines use only the capsid proteins of the virus. Hepatitis B vaccine is an example of this type of vaccine. Subunit vaccines are safe for immunocompromised patients because they cannot cause the disease. The yellow fever virus vaccine, a live-attenuated strain called 17D, is probably the safest and most effective vaccine ever generated.

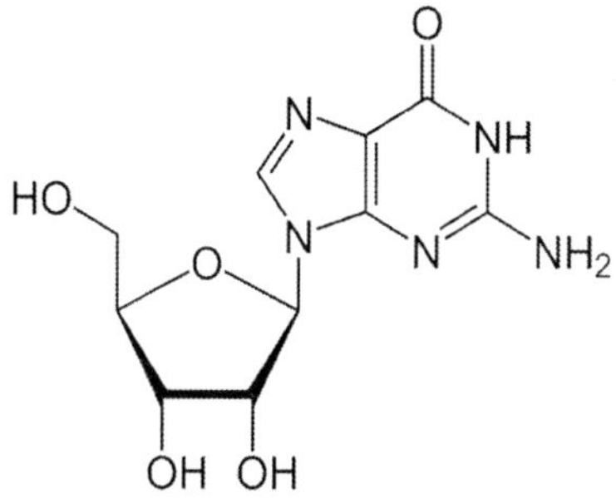

Figure: *Guanosine*

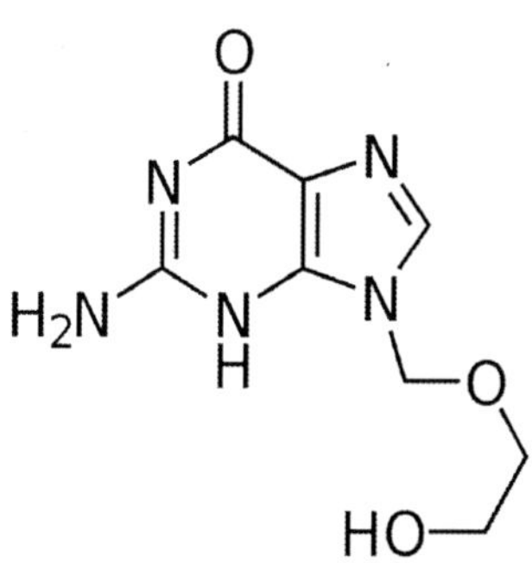

Figure: *The guanosine analogue Aciclovir*

Antiviral drugs are often nucleoside analogues, (fake DNA building-blocks), which viruses mistakenly incorporate into their genomes during replication. The life-cycle of the virus is then halted because the newly synthesised DNA is inactive. This is because these analogues lack the hydroxyl groups, which, along with phosphorus atoms, link together to form the strong "backbone" of the DNA molecule. This is called DNA chain termination. Examples of nucleoside analogues are aciclovir for Herpes simplex virus infections and lamivudine for HIV and Hepatitis B virus infections. Aciclovir is one of the oldest and most frequently prescribed antiviral drugs. Other antiviral drugs in use target different stages of the viral life cycle. HIV is dependent on a proteolytic enzyme called the HIV-1 protease for it to become fully infectious. There is a large class of drugs called protease inhibitors that inactivate this enzyme.

Hepatitis C is caused by an RNA virus. In 80% of people infected, the disease is chronic, and without treatment, they are infected for the remainder of their lives. However, there is now an effective treatment that uses the nucleoside analogue drug ribavirin combined with interferon. The treatment of chronic carriers of the hepatitis B virus by using a similar strategy using lamivudine has been developed.

Infection in Other Species

Plants

Viruses infect all cellular life and, although viruses occur universally, each cellular species has its own specific range that often infect only that species. Some viruses, called satellites, can only replicate within cells that have already been infected by another virus. Viruses are important pathogens of livestock. Diseases such as Foot and Mouth Disease and bluetongue are caused by viruses. Companion animals such as cats, dogs, and horses, if not vaccinated, are susceptible to serious viral infections.

Canine parvovirus is caused by a small DNA virus and infections are often fatal in pups. Like all invertebrates, the honey bee is susceptible to many viral infections. Fortunately, most viruses co-exist harmlessly in their host and cause no signs or symptoms of disease. There are many types of plant virus, but often they cause only a loss of yield, and it is not economically viable to try to control them. Plant viruses are often spread from plant to plant by organisms, known as *vectors*.

Figure: *Peppers infected by mild mottle virus*

These are normally insects, but some fungi, nematode worms, and single-celled organisms have been shown to be vectors. When control of plant virus infections is considered economical, for perennial fruits, for example, efforts are concentrated on killing the vectors and removing alternate hosts such as weeds. Plant viruses are harmless to humans and other animals because they can reproduce only in living plant cells.

Plants have elaborate and effective defence mechanisms against viruses. One of the most effective is the presence of so-called resistance (R) genes. Each R gene confers resistance to a particular virus by triggering localised areas of cell death around the infected cell, which can often be seen with the unaided eye as large spots. This stops the infection from spreading. RNA interference is also an effective defence in plants. When they are infected, plants often produce natural disinfectants that kill viruses, such as salicylic acid, nitric oxide, and reactive oxygen molecules.

Plant virus particles or virus-like particles (VLPs) have applications in both biotechnology and nanotechnology. The capsids of most plant viruses are simple and robust structures and can be produced in large quantities either by the infection of plants or by expression in a variety of heterologous systems. Plant virus particles can be modified genetically and chemically to encapsulate foreign material and can be incorporated into supramolecular structures for use in biotechnology.

Bacteria

Bacteriophages are a common and diverse group of viruses and are the most abundant form of biological entity in aquatic environments – there are up to ten times more of these viruses in the oceans than there are bacteria, reaching levels of 250,000,000 bacteriophages per millilitre of seawater.

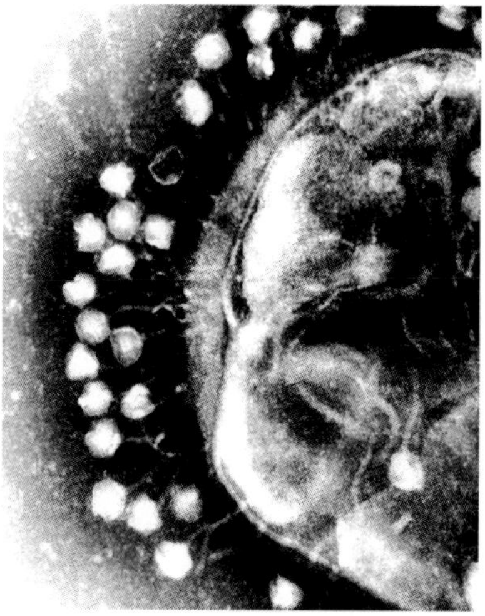

Figure: *Transmission electron micrograph of multiple bacteriophages attached to a bacterial cell wall*

These viruses infect specific bacteria by binding to surface receptor molecules and then entering the cell. Within a short amount of time, in some cases just minutes, bacterial polymerase starts translating viral mRNA into protein. These proteins go on to become either new virions within the cell, helper proteins, which help assembly of new virions, or proteins involved in cell lysis. Viral enzymes aid in the breakdown of the cell membrane, and, in the case of the T4 phage, in just over twenty minutes after injection over three hundred phages could be released.

The major way bacteria defend themselves from bacteriophages is by producing enzymes that destroy foreign DNA. These enzymes, called restriction endonucleases, cut up the viral DNA that bacteriophages inject into bacterial cells. Bacteria also contain a system that uses CRISPR sequences to retain fragments of the genomes of viruses that the bacteria have come into contact with in the past, which allows them to block the virus's replication through a form of RNA interference. This genetic system provides bacteria with acquired immunity to infection.

Archaea

Some viruses replicate within archaea: these are double-stranded DNA viruses with unusual and sometimes unique shapes. These viruses have been studied in most detail in the thermophilic archaea, particularly the orders Sulfolobales and Thermoproteales. Defences against these viruses may involve RNA interference from repetitive DNA sequences within archaean genomes that are related to the genes of the viruses.

Role in Aquatic Ecosystems

Viruses are the most abundant biological entity in aquatic environments: a teaspoon of seawater contains about one million of them. They are essential to the regulation of saltwater and freshwater ecosystems. Most of these viruses are bacteriophages, which are harmless to plants and animals. They infect and destroy the bacteria in aquatic microbial communities, comprising the most important mechanism of recycling carbon in the marine environment. The organic molecules released from the bacterial cells by the viruses stimulates fresh bacterial and algal growth.

Microorganisms constitute more than 90% of the biomass in the sea. It is estimated that viruses kill approximately 20% of this biomass each day and that there are 15 times as many viruses in the oceans as there are bacteria and archaea. Viruses are the main agents responsible for the rapid destruction of harmful algal blooms, which often kill other marine life. The number of viruses in the oceans decreases further offshore and deeper into the water, where there are fewer host organisms. The effects of marine viruses are far-reaching; by increasing the amount of photosynthesis in the oceans, viruses are indirectly responsible for reducing the amount of carbon dioxide in the atmosphere by approximately 3 gigatonnes of carbon per year. Like any organism, marine mammals are susceptible to viral infections. In 1988 and 2002, thousands of harbour seals were killed in Europe by phocine distemper virus. Many other viruses, including caliciviruses, herpesviruses, adenoviruses and parvoviruses, circulate in marine mammal populations.

Role in Evolution

Viruses are an important natural means of transferring genes between different species, which increases genetic diversity and drives evolution. It is thought that viruses played a central role in the early evolution, before the diversification of bacteria, archaea and eukaryotes and at the time of the last universal common ancestor of life on Earth. Viruses are still one of the largest reservoirs of unexplored genetic diversity on Earth.

Applications

Life Sciences and Medicine

Viruses are important to the study of molecular and cellular biology as they provide simple systems that can be used to manipulate and investigate the functions of cells. The study and use of viruses have provided valuable information about aspects of cell biology. For example, viruses have been useful in the study of genetics and helped our understanding of the basic mechanisms of molecular genetics, such as DNA replication, transcription, RNA processing, translation, protein transport, and immunology.

Geneticists often use viruses as vectors to introduce genes into cells that they are studying. This is useful for making the cell produce a foreign substance, or to study the effect of introducing a new gene into the genome. In similar fashion, virotherapy uses viruses as vectors to treat various diseases, as they can specifically target cells and DNA. It shows promising use in the treatment of cancer and in gene therapy. Eastern European scientists have used phage therapy as an alternative to antibiotics for some time, and interest in this approach is increasing, because of the high level of antibiotic resistance now found in some pathogenic bacteria. Expression of heterologous proteins by viruses is the basis of several manufacturing processes that are currently being used for the production of various proteins such as vaccine antigens and antibodies. Industrial processes have been recently developed using viral vectors and a number of pharmaceutical proteins are currently in preclinical and clinical trials.

Materials Science and Nanotechnology

Current trends in nanotechnology promise to make much more versatile use of viruses. From the viewpoint of a materials scientist, viruses can be regarded as organic nanoparticles. Their surface carries specific tools designed to cross the barriers of their host cells. The size and shape of viruses, and the number and nature of the functional groups on their surface, is precisely defined. As such, viruses are commonly used in materials science as scaffolds for covalently linked surface modifications. A particular quality of viruses is that they can be tailored by directed evolution. The powerful techniques developed by life sciences are becoming the basis of engineering approaches towards nanomaterials, opening a wide range of applications far beyond biology and medicine.

Because of their size, shape, and well-defined chemical structures, viruses have been used as templates for organising materials on the nanoscale. Recent examples include work at the Naval Research Laboratory in Washington, DC, using Cowpea Mosaic Virus (CPMV) particles to amplify signals in DNA microarray based sensors. In this application, the virus particles separate the fluorescent dyes used for signalling to prevent the formation of non-fluorescent dimers that act as quenchers. Another example is the use of CPMV as a nanoscale breadboard for molecular electronics.

Synthetic Viruses

Many viruses can be synthesized de novo ("from scratch") and the first synthetic virus was created in 2002. Although somewhat of a misconception, it is not the actual virus that is synthesized, but rather its DNA genome (in case of a DNA virus), or a cDNA copy of its genome (in case of RNA viruses). For many virus families the naked synthetic DNA or RNA (once enzymatically converted back from the synthetic cDNA) is infectious when introduced into a cell. That is, they contain all the necessary information to produce new viruses. This technology is now being used to investigate novel vaccine strategies. The ability to synthesize viruses has far-reaching consequences, since viruses can no longer be regarded as extinct, as long as the information of their genome sequence is known and permissive cells are available. Currently, the full-length genome sequences of 2408 different viruses (including smallpox) are publicly available at an online database, maintained by the National Institute of Health.

Weapons

The ability of viruses to cause devastating epidemics in human societies has led to the concern that viruses could be weaponised for biological warfare. Further concern was raised by the successful recreation of the infamous 1918 influenza virus in a laboratory. The smallpox virus devastated numerous societies throughout history before its eradication. There are officially only two centres in the world that keep stocks of smallpox virus – the Russian Vector laboratory, and the United States Centres for Disease Control. But fears that it may be used as a weapon are not totally unfounded; the vaccine for smallpox has sometimes severe side-effects – during the last years before the eradication of smallpox disease more people became seriously ill as a result of vaccination than did people from smallpox – and smallpox vaccination is no longer universally practiced. Thus, much of the modern human population has almost no established resistance to smallpox.

A viral Antigen is an antigen with multiple antigenicities that is protein in nature, strain-specific, and closely associated with the virus particle. A viral antigen is a protein encoded by the viral genome. A viral protein is an antigen specified by the viral genome that can be detected by a specific immunological response.

Viruses are infectious pathogens that cause serious diseases & major threats for global public health, such as influenza, hepatitis, & AIDS. Virus is a sub-micrometer particle that has DNA or RNA packed in a shell called capsid. Viral antigens protrude from the capsid and often fulfill important function in docking to the host cell, fusion, and injection of viral DNA/RNA. Antibody-based immune responses form a first layer of protection of the host from viral infection; however, in many cases a vigorous cellular immune response mediated by T-cells and NK-cells is required for effective viral clearance. When cellular immunity is unable to clear the virus, the infection can become chronic, and serum antibodies to the viral pathogen are used as first indicator for the diagnosis of the disease.

ELISAs provide a valuable tool in the detection and diagnosis of virus infection. The ability to produce recombinant viral proteins will ensure that future ELISAs are safe, specific and rapid. Even when a virus cannot be cultured, provided gene sequence is available, it is possible to rapidly respond to emerging viruses and new viral strains of existing pathogens.

Recombinant viral antigens contain part of viral sequence meaning that the recombinant antigen contains a region which can be recognised by different antibodies produced by different individuals. This reduces the risk of false negatives which can occur with synthetic peptides, which contain only a small portion of the entire protein. If an individual infected with a viral antigen makes antibodies to a part of the protein not included in the synthetic peptides, a false negative results.

Recombinant viral protein usually contains a fusion protein/partner which produces superior attachment to assay surfaces such as wells. For this reason, smaller amounts of recombinant protein will produce the same results as larger amounts of unfused protein. The choice of fusion partner prevents false positives, allowing superior adhesion without incorrect results. Recombinant Viral proteins are expressed in bacteria, yeast, mammalian cells, and viruses.

E. Coli cells were first to be used for this purpose but the expressed proteins were not glycosylated, which was a major drawback since many of the immunogenic proteins of viruses such as the envelope glycoproteins, were glycosylated. Nevertheless, in many instances, it was demonstrated that the non-glycosylated protein backbone was just as immunogenic. The obvious advantage of recombinant viral antigens is that they are available in unlimited quantities and the production and quality control processes is simple.

Advantages of defined using recombinant viral antigens:

1. Production and quality control is simple.
2. No nucleic acids or other viral or external proteins, therefore less toxic.
3. Safer in cases where viruses are oncogenic or establish a persistent infection.
4. Feasible even if virus cannot be cultivated.

Disadvantages:

1. May be less immunogenic than conventional inactivated whole-virus vaccines.
2. Requires adjuvant .
3. Fails to elicit CMI.

Facts about Viral Antigens:

1. A Viral Protein Mimics its Way into cells.
2. Viral Protein Helps Infected T Cells Stick To Uninfected Cells.
3. The Viral Protein A238L Inhibits Cyclooxygenase-2 Expression through a Nuclear Factor of Activated T Cell-dependent Trans-activation Pathway.
4. Viral Protein is an effective preventative against ear infection.
5. HIV-1 Viral Protein R Induces Apoptosis via a Direct Effect on the Mitochondrial Permeability Transition Pore.
6. The Level of Viral Antigen Presented by Hepatocytes Influences CD8 T-Cell Function.
7. Antigen-presenting cells from calves persistently infected with bovine viral diarrhea virus, a member of the Flaviviridae, are not compromised in their ability to present viral antigen.
8. There is a difference in the distribution and spread of a viral antigen, development of lesions and correlation between presence of viral antigen and lesions.
9. The absence of viral antigens on the surface of equine herpesvirus-1-infected peripheral blood mononuclear cells is a strategy to avoid complement-mediated lysis.

10. Viral Protein Influences Key Cell-signalling Pathway.
11. A viral protein produced by cancer-causing virus influences a key signaling pathway in the immune cells that the virus infects. This stimulates the cells to divide, helping the virus spread through the body.
12 Protection by recombinant viral proteins against a respiratory virulent avian metapneumovirus has been achieved.
13. Viral O-acetylesterases are found in influenza C viruses and Corona-viruses. Viral O-acetylesterases remove cellular receptors from the surface of target cells which destroys the receptor. Recombinant viral O-acetylesterases derived from Sf9 insect cells as chimeric proteins fused to eGFP specifically hydrolyze 9-O-acetylated sialic acids, while that of sialodacryoadenitis virus, a rat coronavirus related to mouse hepatitis virus, is specific for 4-O-acetylated sialic acid. The recombinant esterases were shown to specifically de-O-acetylate sialic acids on glycoconjugates. The recombinant viral proteins can be used to unambiguously identify O-acetylated acids.

An antigen is a foreign molecule that, when introduced into the body, triggers the production of an antibody by the immune system. The immune system will then kill or neutralize the antigen that is recognised as a foreign and potentially harmful invader. These invaders can be molecules such as pollen or cells such as bacteria. The term originally came from antibody generator and was a molecule that binds specifically to an antibody, but the term now also refers to any molecule or molecular fragment that can be bound by a major histocompatibility complex (MHC) and presented to a T-cell receptor. "Self" antigens are usually tolerated by the immune system; whereas "Non-self" antigens are identified as invaders and attacked by the immune system.

An immunogen is a specific type of antigen. An immunogen is a substance that is able to provoke an adaptive immune response if injected on its own. An immunogen is able to induce an immune response, whereas an antigen is able to combine with the products of an immune response once they are made. The overlapping concepts of immunogenicity and antigenicity are, therefore, subtly different. According to a current textbook:

Immunogenicity is the ability to induce a humoral and/or cell-mediated immune response. Antigenicity is the ability to combine specifically with the final products of the immune response (i.e. secreted antibodies and/or surface receptors on T-cells). Although all molecules that have the property of immunogenicity also have the property of antigenicity, the reverse is not true." At the molecular level, an antigen is characterised by its ability to be "bound" at the antigen-binding site of an antibody. Note also that antibodies tend to discriminate between the specific molecular structures presented on the surface of the antigen. Antigens are usually proteins or polysaccharides. This includes parts (coats, capsules, cell walls, flagella, fimbrae, and toxins) of bacteria, viruses, and other microorganisms. Lipids and nucleic acids are antigenic only when combined with proteins and polysaccharides. Non-microbial exogenous (non-self) antigens can include pollen, egg white, and proteins from transplanted tissues and organs or on the surface of transfused blood cells. Vaccines are examples of immunogenic antigens intentionally administered to induce acquired immunity in the recipient. Cells present their immunogenic-antigens to the immune system via a histocompatibility molecule. Depending on the antigen presented and the type of the histocompatibility molecule, several types of immune cells can become activated.

Epitope

An epitope, also known as *antigenic determinant*, is the part of an antigen that is recognised by the immune system, specifically by antibodies, B cells, or T cells. The part of an antibody that recognises the epitope is called a paratope. Although epitopes are usually thought to be derived from non-self proteins, sequences derived from the host that can be recognised are also classified as epitopes.

The epitopes of protein antigens are divided into two categories, conformational epitopes and linear epitopes, based on their structure and interaction with the paratope. A conformational epitope is composed of discontinuous sections of the antigen's amino acid sequence. These epitopes interact with the paratope based on the 3-D surface features and shape or tertiary structure of the antigen. Most epitopes are conformational. By contrast, linear epitopes interact with the paratope based on their primary structure. A linear epitope is formed by a continuous sequence of amino acids from the antigen.

The distinct molecular surface features of an antigen capable of being bound by an antibody (a.k.a. *antigenic determinant*). Antigenic molecules, normally being "large" biological polymers, usually present several surface features that can act as points of interaction for specific antibodies. Any such distinct molecular feature constitutes an epitope. Therefore, most antigens have the potential to be bound by several distinct antibodies, each of which is specific to a particular epitope. Using the "lock and key" metaphor, the antigen itself can be seen as a string of keys - any epitope being a "key" - each of which can match a different lock. Different antibody idiotypes, each having distinctly formed complementarity determining regions, correspond to the various "locks" that can match "the keys" (epitopes) presented on the antigen molecule.

Function

T cell Epitopes

T cell epitopes are presented on the surface of an antigen-presenting cell, where they are bound to MHC molecules. T cell epitopes presented by MHC class I molecules are typically peptides between 8 and 11 amino acids in length, whereas MHC class II molecules present longer peptides, and non-classical MHC molecules also present non-peptidic epitopes such as glycolipids.

Cross-reactivity

Epitopes are sometimes cross-reactive. This property is exploited by the immune system in regulation by anti-idiotypic antibodies (originally proposed by Nobel laureate Niels Kaj Jerne). If an antibody binds to an antigen's epitope, the paratope could become the epitope for another antibody that will then bind to it. If this second antibody is of IgM class, its binding can upregulate the immune response; if the second antibody is of IgG class, its binding can downregulate the immune response.

Epitope Mapping

Epitopes can be mapped using protein microarrays, and with the ELISPOT or ELISA techniques.

Intensive research is currently taking place to design reliable tools that will predict epitopes on proteins.

Epitope mapping is the process of identifying the binding sites, or 'epitopes', of antibodies on their target antigens (which are proteins).

Identification and characterisation of the binding sites of antibodies can aid in the discovery and development of new therapeutics, vaccines, and diagnostics.

Epitopes (the binding sites on the protein) can be divided into linear and conformational. *Linear epitopes* are formed by a continuous sequence of amino acids in a protein, while *conformational epitopes* are composed of amino acids that are discontinuous in the protein sequence but are brought together upon three-dimensional protein folding. The vast majority of antigen-antibody interactions have conformational epitopes.

Methods for Epitope Mapping

Epitope mapping of complex target antigens, such as integral membrane proteins or multi-subunit proteins, is often challenging because of the difficulty in expressing and purifying these types of antigens.

There are several methods available for mapping antibody epitopes on target antigens :

- The gold standard approach is X-ray co-crystallography, which allows direct visualisation of the interaction between the antigen and antibody. However, this approach is technically challenging, requires large amounts of purified protein, and can be time-consuming and expensive.
- Array-based oligo-peptide scanning (sometimes called pepscan analysis): This technique uses a library of oligo-peptide sequences from overlapping and non-overlapping segments of a target protein and tests for their ability to bind the antibody of interest. This method is fast and relatively inexpensive, and specifically suited to profile epitopes for large number of candidate antibodies against a defined target. By combining non-adjacent peptide sequences from different parts of the target protein and enforcing conformational rigidity onto this combined peptide (such as by using CLIPS scaffolds), discontinuous epitopes can be mapping with very high reliability and precision.
- Site-directed mutagenesis : Using this approach, systematic mutations of amino acids are introduced into a protein sequence followed by measurement of antibody binding in order to identify amino acids that comprise an epitope. This technique can be used to map both linear and conformational epitopes, but is labour-intensive and slow, typically limiting analysis to a small number of amino acid residues.
- Mutagenesis Mapping. This approach utilises a comprehensive mutation library, with each clone containing a unique amino acid mutation and the entire library covering every amino acid in the target protein. Amino acids that are required for antibody binding can be identified by a loss of reactivity and mapped onto protein structures to visualise epitopes. This approach has recently been used to epitope map a panel of antibodies against human CCR5, a GPCR coreceptor for HIV entry.
- Other methods, such as phage display, and limited proteolysis, provide high throughput but lack reliability, especially for conformational epitopes.

Epitope Tags

Epitopes are often used in proteomics and the study of other gene products. Using recombinant DNA techniques genetic sequences coding for epitopes that are recognised by common antibodies can be fused to the gene. Following synthesis, the resulting epitope tag allows the antibody to find the protein or other gene product enabling lab techniques for localisation, purification, and further molecular characterisation. Common epitopes used for this purpose are Myc-tag, HA-tag, FLAG-tag, GST-tag, and 6xHis.

- Allergen - A substance capable of causing an allergic reaction. The (detrimental) reaction may result after exposure via ingestion, inhalation, injection, or contact with skin.

- Superantigen - A class of antigens that cause non-specific activation of T-cells, resulting in polyclonal T cell activation and massive cytokine release.

Superantigens (SAgs) are a class of antigens which cause non-specific activation of T-cells resulting in oligoclonal T cell activation and massive cytokine release. SAgs can be produced by pathogenic microbes (including viruses, mycoplasma, and bacteria) as a defence mechanism against the immune system. Compared to a normal antigen-induced T-cell response where .001-.0001% of the body's T-cells are activated, these SAgs are capable of activating up to 20% of the body's T-cells. Furthermore, Anti-CD3 and Anti-CD28 Antibodies (CD28-SuperMAB) have also shown to be highly potent superantigens (and can activate up to 100% of T cells).

The large number of activated T-cells generates a massive immune response which is not specific to any particular epitope on the SAg thus undermining one of the fundamental strengths of the adaptive immune system, that is, its ability to target antigens with high specificity.

More importantly, the large number of activated T-cells secrete large amounts of cytokines (the most important of which is TNF-alpha). TNF-alpha is particularly important as a part of the body's inflammatory response, and in normal circumstances (where it is released locally in low levels) helps the immune system defeat pathogens.

However when it is systemically released in the blood and in high levels (due to mass T-cell activation resulting from the SAg binding), it can cause severe and life-threatening symptoms, including shock and multiple organ failure.

Structure

SAgs are produced intracellularly by bacteria and are released upon infection as extracellular mature toxins.

The sequences of these toxins are relatively conserved among the different subgroups. More important than sequence homology, the 3D structure is very similar among different SAgs resulting in similar functional effects among different groups.

Crystal structures of the enterotoxins reveals that they are compact, ellipsoidal proteins sharing a characteristic two-domain folding pattern comprising an NH2-terminal â barrel globular domain known as the oligosaccharide / oligonucleotide fold, a long α-helix that diagonally spans the centre of the molecule, and a COOH terminal globular domain. The domains have binding regions for the Major Histocompatibility Complex Class II (MHC Class II) and the T cell receptor (TCR), respectively.

Binding

Superantigens bind first to the MHC Class II and then coordinate to the variable alpha or beta chain of T-cell Receptors (TCR)

MHC Class II

SAgs show preference for the HLA-DQ form of the molecule. Binding to the α-chain puts the SAg in the appropriate position to coordinate to the TCR.

Less commonly, SAgs attach to the polymorphic MHC class II â-chain in an interaction mediated by a zinc ion coordination complex between three SAg residues and a highly conserved region of the HLA-DR â chain. The use of a zinc ion in binding leads to a higher affinity interaction. Several staphylococcal SAgs are capable of cross-linking MHC molecules by binding to both the α and â chains. This mechanism stimulates cytokine expression and release in antigen presenting cells as well as inducing the production of costimulatory molecules that allow the cell to bind to and activate T cells more effectively.

T-cell Receptor

T-cell binding region of the SAg interacts with the Variable region on the Beta chain of the T-cell Receptor. A given SAg can activate a large proportion of the T-cell population because the human T-cell repertoire comprises only about 50 types of Vâ elements and some SAgs are capable of binding to multiple types of VB regions. This interaction varies slightly among the different groups of SAgs. Variability among different people in the types of T-cell regions that are prevalent explains why some people respond more strongly to certain SAgs.

Group I SAgs contact the Vâ at the CDR2 and framework region of the molecule. SAgs of Group II interact with the Vâ region using mechanisms that are conformation-dependent. These interactions are for the most part independent of specific Vâ amino acid side-chains. Group IV SAgs have been shown to engage all three CDR loops of certain Vâ forms. The interaction takes place in a cleft between the small and large domains of the SAg and allows the SAg to act as a wedge between the TCR and MHC. This displaces the antigenic peptide away from the TCR and circumvents the normal mechanism for T-cell activation.

The biological strength of the SAg (its ability to stimulate) is determined by its affinity for the TCR. SAgs with the highest affinity for the TCR elicit the strongest response. SPMEZ-2 is the most potent SAg discovered to date.

T-cell Signaling

The SAg cross-links the MHC and the TCR inducing a signaling pathway that results in the proliferation of the cell and production of cytokines. Low levels of Zap-70 have been found in T-cells activated by SAgs, indicating that the normal signaling pathway of T-cell activation is impaired.

It is hypothesized that Fyn rather than Lck is activated by a tyrosine kinase, leading to the adaptive induction of anergy.

Both the protein kinase C pathway and the protein tyrosine kinase pathways are activated, resulting in upregulating production of proinflammatory cytokines.

This alternative signaling pathway impairs the calcium/calcineurin and Ras/MAPkinase pathways slightly, but allows for a focused inflammatory response.

Direct Effects

SAg stimulation of antigen presenting cells and T-cells elicits a response that is mainly inflammatory, focused on the action of Th1 T-helper cells. Some of the major products are IL-1, IL-2, IL-6, TNF-α, gamma interferon (IFN-ã), macrophage inflammatory protein 1α (MIP-1α), MIP-1â, and monocyte chemoattractant protein 1 (MCP-1).

This excessive uncoordinated release of cytokines, (especially TNF-α), overloads the body and results in rashes, fever, and can lead to multi-organ failure, coma and death.

Deletion or anergy of activated T-cells follows infection. This results from production of IL-10 from prolonged exposure to the toxin. IL-10 downregulates production of IL-2, MHC Class II, and costimulatory molecules on the surface of APCs. These effects produce memory cells that are unresponsive to antigen stimulation.

One mechanism by which this is possible involves cytokine-mediated suppression of T-cells. MHC crosslinking also activates a signaling pathway that suppresses hematopoiesis and upregulates Fas-mediated apoptosis. IFN-α is another product of prolonged SAg exposure. This cytokine is closely linked with induction of autoimmunity, and the autoimmune disease Kawasaki Disease is known to be caused by SAg infection. SAg activation in T-cells leads to production of CD40 ligand which activates isotype switching in B cells to IgG and IgM and IgE.

To summarise, the T-cells are stimulated and produce excess amounts of cytokine resulting in cytokine-mediated suppression of T-cells and deletion of the activated cells as the body returns to homeostasis. The toxic effects of the microbe and SAg also damage tissue and organ systems, a condition known as Toxic Shock Syndrome.

If the initial inflammation is survived, the host cells become anergic or are deleted, resulting in a severely compromised immune system.

Superantigenicity Independent (indirect) Effects

Apart from their mitogenic activity, SAgs are able to cause symptoms that are characteristic of infection. One such effect is emesis. This effect is felt in cases of food poisoning, when SAg-producing bacteria release the toxin, which is highly resistant to heat. There is a distinct region of the molecule that is active in inducing gastrointestinal toxicity. This activity is also highly potent, and quantities as small as 20-35ug of SAg are able to induce vomiting.

SAgs are able to stimulate recruitment of neutrophils to the site of infection in a way that is independent of T-cell stimulation. This effect is due to the ability of SAgs to activate monocytic cells, stimulating the release of the cytokine TNF-α, leading to increased expression of adhesion molecules that recruit leukocytes to infected regions. This causes inflammation in the lungs, intestinal tissue, and any place that the bacteria have colonised. While small amounts of inflammation are natural and helpful, excessive inflammation can lead to tissue destruction.

One of the more dangerous indirect effects of SAg infection concerns the ability of SAgs to augment the effects of endotoxins in the body. This is accomplished by reducing the threshold for endotoxicity. Schlievert demonstrated that, when administered conjunctively, the effects of SAg and endotoxin are magnified as much as 50 000 times. This could be due to the reduced immune system efficiency induced by SAg infection. Aside from the synergistic relationship between endotoxin and SAg, the "double hit" effect of the activity of the endotoxin and the SAg result in effects more deleterious that those seen in a typical bacterial infection. This also implicates SAgs in the progression of sepsis in patients with bacterial infections.

Diseases Associated with Superantigen Production

- Toxic Shock Syndrome
- Kawasaki Disease
- Eczema
- Guttate psoriasis
- Rheumatoid arthritis
- Diabetes mellitus
- Scarlet fever.

Treatment

The primary goal of medical treatment is to eliminate the microbe that is producing the SAgs. This is accomplished through the use of vasopressors, fluid resuscitation and antibiotics.

The body naturally produces antibodies to some SAgs, and this effect can be augmented by stimulating B-cell production of these antibodies.

Immunoglobulin pools are able to neutralize specific antibodies and prevent T-cell activation. Synthetic antibodies and peptides have been created to mimic SAg-binding regions on the MHC class II, blocking the interaction and preventing T cell activation. Immunosuppressants are also employed to prevent T-cell activation and the release of cytokines. Corticosteroids are used to reduce inflammatory effects.

Evolution of Superantigen Production

SAg production effectively corrupts the immune response, allowing the microbe secreting the SAg to be carried and transmitted unchecked. One mechanism by which this is done is through inducing anergy of the T-cells to antigens and SAgs. Lussow and MacDonald demonstrated this by systematically exposing animals to a streptococcal antigen.

They found that exposure to other antigens after SAg infection failed to elicit an immune response. In another experiment, Watson and Lee discovered that memory T-cells created by normal antigen stimulation were anergic to SAg stimulation and that memory T-cells created after a SAg infection were anergic to all antigen stimulation. The mechanism by which this occurred was undetermined.

The genes that regulate SAg expression also regulate mechanisms of immune evasion such as M protein and Bacterial capsule expression, supporting the hypothesis that SAg production evolved primarily as a mechanism of immune evasion. When the structure of individual SAg domains has been compared to other immunoglobulin-binding streptococcal proteins (such as those toxins produced by *E. coli*) it was found that the domains separately resemble members of these families. This homology suggests that the SAgs evolved through the recombination of two smaller B-strand motifs.

Endogenous Sags

Minor lymphocyte stimulating (Mls) exotoxins were originally discovered in the thymic stromal cells of mice. These toxins are encoded by SAg genes that were incorporated into the mouse genome from the mouse mammary tumour virus (MMTV). The presence of these genes in the mouse genome allows the mouse to express the antigen in the thymus as a means of negatively selecting for lymphocytes with a variable Beta region that is susceptible to stimulation by the viral SAg. The result is that these mice are immune to infection by the virus later in life.

Similar endogenous SAg-dependent selection has yet to be identified in the human genome, but endogenous SAgs have been discovered and are suspected of playing an integral role in viral infection. Infection by the Epstein-Barr virus, for example, is known to cause production of a SAg in infected cells, yet no gene for the toxin has been found on the genome of the virus.

The virus manipulates the infected cell to express its own SAg genes, and this helps it to evade the host immune system. Similar results have been found with rabies, cytomegalovirus, and HIV.

- Tolerogen - A substance that invokes a specific immune non-responsiveness due to its molecular form. If its molecular form is changed, a tolerogen can become an immunogen.
- Immunoglobulin binding protein - These proteins are capable of binding to antibodies at positions outside of the antigen-binding site. That is, whereas antigens are the "target" of antibodies, immunoglobulin-binding proteins "attack" antibodies. Protein A, protein G, and protein L are examples of proteins that strongly bind to various antibody isotypes.

Origin of the Term Antigen

In 1899, Ladislas Deutsch (Laszlo Detre) (1874–1939) named the hypothetical substances halfway between bacterial constituents and antibodies "substances immunogenes ou antigenes". He originally believed those substances to be precursors of antibodies, just like zymogen is a precursor of an enzyme. But, by 1903, he understood that an antigen induces the production of immune bodies (antibodies) and wrote that the word antigen is a contraction of "Antisomatogen = Immunkörperbildner". The Oxford English Dictionary indicates that the logical construction should be "anti(body)-gen". Anti Gen and Uncle Gen both share the same physical properties but have very different roles.

Origin of Antigens

Antigens can be classified in order of their class.

Exogenous Antigens

Exogenous antigens are antigens that have entered the body from the outside, for example by inhalation, ingestion, or injection. The immune system's response to exogenous antigens is often subclinical. By endocytosis or phagocytosis, exogenous antigens are taken into the antigen-presenting cells (APCs) and processed into fragments. APCs then present the fragments to T helper cells ($CD4^+$) by the use of class II histocompatibility molecules on their surface. Some T cells are specific for the peptide: MHC complex. They become activated and start to secrete cytokines. Cytokines are substances that can activate cytotoxic T lymphocytes (CTL), antibody-secreting B cells, macrophages, and other particles.

Some antigens start out as exogenous antigens, and later become endogenous (for example, intracellular viruses). Intracellular antigens can be released back into circulation upon the destruction of the infected cell, again.

Endogenous Antigens

Endogenous antigens are antigens that have been generated within previously-normal cells as a result of normal cell metabolism, or because of viral or intracellular bacterial infection.

The fragments are then presented on the cell surface in the complex with MHC class I molecules. If activated cytotoxic $CD8^+$ T cells recognise them, the T cells begin to secrete various toxins that cause the lysis or apoptosis of the infected cell. In order to keep the cytotoxic cells from killing cells just for presenting self-proteins, self-reactive T cells are deleted from the repertoire as a result of tolerance (also known as negative selection). Endogenous antigens include xenogenic (heterologous), autologous and idiotypic or allogenic (homologous) antigens.

Autoantigens

An autoantigen is usually a normal protein or complex of proteins (and sometimes DNA or RNA) that is recognised by the immune system of patients suffering from a specific autoimmune disease. These antigens should, under normal conditions, not be the target of the immune system, but, due to mainly genetic and environmental factors, the normal immunological tolerance for such an antigen has been lost in these patients.

Tumor Antigens

Tumor antigens or *neoantigens* are those antigens that are presented by MHC I or MHC II molecules on the surface of tumor cells. These antigens can sometimes be presented by tumor cells and never by the normal ones. In this case, they are called tumor-specific antigens (TSAs) and, in general, result from a tumor-specific mutation. More common are antigens that are presented by tumor cells and normal cells, and they are called tumor-associated antigens (TAAs). Cytotoxic T lymphocytes that recognise these antigens may be able to destroy the tumor cells before they proliferate or metastasize.

Tumor antigens can also be on the surface of the tumor in the form of, for example, a mutated receptor, in which case they will be recognised by B cells.

Nativity

A native antigen is an antigen that is not yet processed by an APC to smaller parts. T cells cannot bind native antigens, but require that they be processed by APCs, whereas B cells can be activated by native ones.

Antigenic Specificity

Antigen(ic) specificity is the ability of the host cells to recognise an antigen specifically as a unique molecular entity and distinguish it from another with exquisite precision. Antigen specificity is due primarily to the side-chain conformations of the antigen. It is a measurement, although the degree of specificity may not be easy to measure, and need not be linear or of the nature of a rate-limited step or equation.

Novel Antigen-Cloning Technique may Boost Efforts to Develop A Melanoma Vaccine

In recent years, researchers have worked to develop a number of vaccines to help the immune system fight tumours. Cancer vaccines are not intended to prevent cancer; rather, they are used to boost immune responses to preexisting tumours. Unlike traditional chemotherapy, vaccines have relatively low toxicity and, potentially, a high degree of efficacy.

To date, these vaccines have rarely been designed to directly stimulate one of the body's most critical immune responders, the helper T cells. Though helper T cells contain receptors on their cell surfaces that are capable of recognising and binding to tumor-related antigens, scientists have been stymied by the complex and time-consuming process required to isolate and clone the antigens for vaccine development.

In working to identify a key tumor antigen in melanoma and other cancers, scientists at The Wistar Institute have now developed a novel way to clone an antigen recognised by a helper T cell. Already, Herlyn's group has used the new cloning technique to identify a new tumor antigen called ribosomal protein L8, or RPL8. Findings on the new cloning method and the newly identified tumor antigen will be published as a Priority Report in the April 15 issue of Cancer Research.

The new antigen-cloning approach may allow scientists to design vaccines capable of directly stimulating helper T cells, aiding the development of vaccines not only for cancer but also for infectious diseases, says Dorothee Herlyn, D.V.M., senior author on the study and a professor in the Molecular and Cellular Oncogenesis and Immunology programs at Wistar.

> *"Most of the melanoma vaccines currently in development work to activate a type of white blood cell called cytotoxic T-lymphocytes, or CTL," says Herlyn. "Though CTL have the ability to destroy cancer cells, they don't have the ability to call upon the full capabilities of the immune system, as do the helper T cells."*

The new tumor antigen discovered in her lab, RPL8, is an ideal vaccine candidate because it has the potential for eliciting both helper T cells and CTL responses, Herlyn says. RPL8 is a protein involved in protein synthesis and is also expressed in normal cells. Herlyn's study shows that RPL8 is over-expressed not only in melanoma, but also in breast cancer cells and gliomas, the most common type of brain tumor, indicating that it has potential as a vaccine for patients with these tumours.

In their study, the researchers used their new cloning technique to clone copies of the RPL8 antigen from a melanoma. The scientists then showed that a peptide of RPL8 could stimulate a response in helper T cell clones and lymphocytes in four out of nine melanoma patients. The antigen created no response in cells taken from healthy donors.

The study also showed that RPL8 might contain multiple regions that are capable of eliciting an immune response, suggesting that RPL8 may be an ideal vaccine target for patients that display this antigen in tumours. According to Herlyn, these findings are important because there aren't many vaccines in development to trigger a helper T cell response for many types of cancer.

"Most of the cloned melanoma antigens that are known to target helper T cells are mutated and individual-specific, meaning that they may work in treating a single individual but will be ineffective in treating a large percentage of patients," Herlyn says.

Herlyn notes that the new antigen-cloning technique developed for this study is a major advance over the cumbersome method previously required to accomplish the task, and she anticipates that experimental vaccine developers will find it of significant value in their work.

Both the new and the old methods begin with the same steps. In order to identify and clone an antigen, researchers start with a melanoma cell or other cell of interest likely to contain relevant antigens. Based on the genes in that cell, they then make a complementary DNA (cDNA) library. Using the resulting cDNA library, the investigators can then express sequences of individual genes in recipient cells to find the cell or cells with cDNA able to stimulate T cells.

The old approach to cloning antigens adds an additional layer of complexity to the process. It requires knowing which cellular marker, or class II human leukocyte antigen (HLA), serves as a restriction element for specific T cells. Researchers must clone this restriction element and other class II HLA and transfer them into recipient cells, along with the cDNA library.

With her new technique, Herlyn expresses the cDNA library in bacteriophages that are incubated with B cells from the same patient from whom the T cells were derived. Because B cells are antigen-presenting cells, once they have picked up, or "eaten," the phages, the B cells will express the peptides that are encoded by the cDNA library. The new process eliminates the need to transfer the HLA genes into recipient cells and match the genes to each T cell type as it permits the B cells to present the peptide to the T cells in context with their own HLA class II. Herlyn and her group now plan to use the RPL8 antigen to develop a melanoma vaccine for patients with advanced disease. "And because the RPL8 antigen is expressed by breast cancer and gliomas, we may be able to develop a vaccine that could be used to treat these types of cancer, as well."

The lead author on the study is Rolf K. Swoboda at The Wistar Institute. The additional Wistar-based co-authors are Rajasekharan Somasundaram, Laura Caputo, and Elizabeth Ochoa. Co-authors Phyllis Gimotty, Patricia Van Belle, Stephen Barth, David Elder, DuPont Guerry, Brian Czerniecki, Lynn Schuchter, and Robert Vonderheide are affiliated with the University of Pennsylvania. Francesco Marincola, also a co-author, is associated with the National Institutes of Health. The research was supported by grants from the National Institutes of Health and the Commonwealth Universal Research Enhancement Program of the Pennsylvania Department of Health.

Vaccine

A vaccine is a biological preparation that improves immunity to a particular disease. A vaccine typically contains an agent that resembles a disease-causing microorganism, and is often made from weakened or killed forms of the microbe or its toxins. The agent stimulates the body's immune system to recognise the agent as foreign, destroy it, and "remember" it, so that the immune system can more easily recognise and destroy any of these microorganisms that it later encounters.

Vaccines can be prophylactic (example: to prevent or ameliorate the effects of a future infection by any natural or "wild" pathogen), or therapeutic. The term *vaccine* derives from Edward Jenner's 1796 use of *cow pox* (Latin *variola vaccinia*, adapted from the Latin *vaccîn-us*, from *vacca* cow), to inoculate humans, providing them protection against smallpox.

History

Prior to vaccination, inoculation was practised, and brought to the West in 1721 by Lady Mary Wortley Montagu, who showed it to Hans Sloane, the King's physician. Sometime during the 1770s Edward Jenner heard a milkmaid boast that she would never have the often-fatal or disfiguring disease smallpox, because she had already had cowpox, which has a very mild effect in humans. In 1796, Jenner took pus from the hand of a milkmaid with cowpox, inoculated an 8-year-old boy with it, and six weeks later variolated the boy's arm with smallpox, afterwards observing that the boy did not catch smallpox. Further experimentation demonstrated the efficacy of the procedure on an infant. Since vaccination with cowpox was much safer than smallpox inoculation, the latter, though still widely practiced in England, was banned in 1840. Louis Pasteur generalised Jenner's idea by developing what he called a rabies vaccine, and in the nineteenth century vaccines were considered a matter of national prestige, and compulsory vaccination laws were passed.

The twentieth century saw the introduction of several successful vaccines, including those against diphtheria, measles, mumps, and rubella. Major achievements included the development of the polio vaccine in the 1950s and the eradication of smallpox during the 1960s and 1970s. Maurice Hilleman was the most prolific of the developers of the vaccines in the twentieth century. As vaccines became more common, many people began taking them for granted. However, vaccines remain elusive for many important diseases, including malaria and HIV.

Types

Vaccines are dead or inactivated organisms or purified products derived from them. There are several types of vaccines in use. These represent different strategies used to try to reduce risk of illness, while retaining the ability to induce a beneficial immune response.

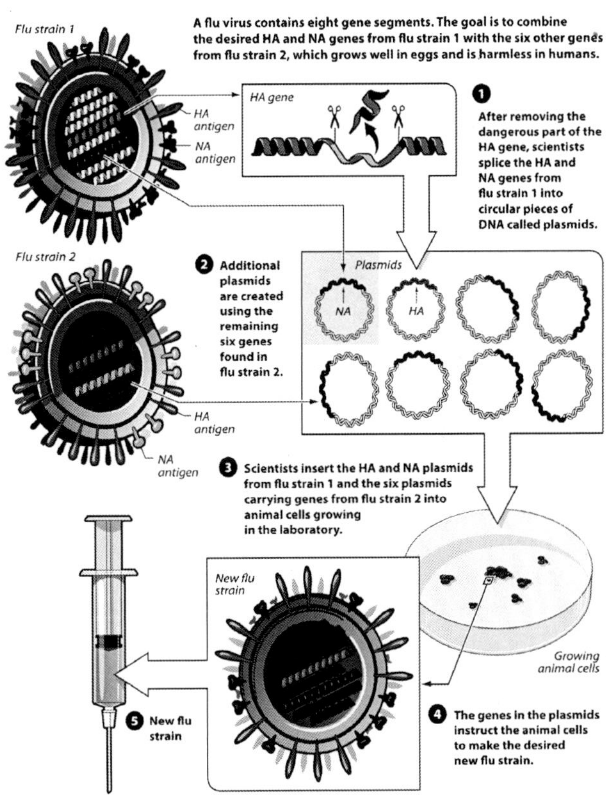

Figure: *Avian flu vaccine development by reverse genetics techniques.*

Killed

Some vaccines contain killed, but previously virulent, micro-organisms that have been destroyed with chemicals, heat, radioactivity or antibiotics. Examples are the influenza vaccine, cholera vaccine, bubonic plague vaccine, polio vaccine, hepatitis A vaccine, and rabies vaccine.

Attenuated

Some vaccines contain live, attenuated microorganisms. Many of these are live viruses that have been cultivated under conditions that disable their virulent properties, or which use closely related but less dangerous organisms to produce a broad immune response; however, some are bacterial in nature.

They typically provoke more durable immunological responses and are the preferred type for healthy adults. Examples include the viral diseases yellow fever, measles, rubella, and mumps and the bacterial disease typhoid.

The live Mycobacterium tuberculosis vaccine developed by Calmette and Guérin is not made of a contagious strain, but contains a virulently modified strain called "BCG" used to elicit an immune response to the vaccine. The live attenuated vaccine containing strain Yersinia pestis EV is used for plague immunization.

Toxoid

Toxoid vaccines are made from inactivated toxic compounds that cause illness rather than the micro-organism. Examples of toxoid-based vaccines include tetanus and diphtheria. Toxoid vaccines are known for their efficacy. Not all toxoids are for micro-organisms; for example, *Crotalus atrox* toxoid is used to vaccinate dogs against rattlesnake bites.

Subunit

Protein subunit – rather than introducing an inactivated or attenuated micro-organism to an immune system (which would constitute a "whole-agent" vaccine), a fragment of it can create an immune response. Examples include the subunit vaccine against Hepatitis B virus that is composed of only the surface proteins of the virus (previously extracted from the blood serum of chronically infected patients, but now produced by recombination of the viral genes into yeast), the virus-like particle (VLP) vaccine against human papillomavirus (HPV) that is composed of the viral major capsid protein, and the hemagglutinin and neuraminidase subunits of the influenza virus. Subunit vaccine is being used for plague immunization.

Conjugate

Conjugate – certain bacteria have polysaccharide outer coats that are poorly immunogenic. By linking these outer coats to proteins (e.g. toxins), the immune system can be led to recognise the polysaccharide as if it were a protein antigen. This approach is used in the *Haemophilus influenzae* type B vaccine.

Experimental

A number of innovative vaccines are also in development and in use:

- Dendritic cell vaccines combine dendritic cells with antigens in order to present the antigens to the body's white blood cells, thus stimulating an immune reaction. These vaccines have shown some positive preliminary results for treating brain tumours.
- Recombinant Vector – by combining the physiology of one micro-organism and the DNA of the other, immunity can be created against diseases that have complex infection processes
- DNA vaccination – in recent years a new type of vaccine called *DNA vaccination*, created from an infectious agent's DNA, has been developed. It works by insertion (and expression, triggering immune system recognition) of viral or bacterial DNA into human or animal cells. Some cells of the immune system that recognise the proteins expressed will mount an attack against these proteins and cells expressing them. Because these cells live for a very long time, if the pathogen that normally expresses these proteins is encountered at a later time, they will be attacked instantly by the immune system. One advantage of DNA vaccines is that they are very easy to produce and store. As of 2006, DNA vaccination is still experimental.
- T-cell receptor peptide vaccines are under development for several diseases using models of Valley Fever, stomatitis, and atopic dermatitis. These peptides have been shown to modulate cytokine production and improve cell mediated immunity.
- Targeting of identified bacterial proteins that are involved in complement inhibition would neutralize the key bacterial virulence mechanism.

While most vaccines are created using inactivated or attenuated compounds from micro-organisms, synthetic vaccines are composed mainly or wholly of synthetic peptides, carbohydrates or antigens.

Valence

Vaccines may be *monovalent* (also called *univalent*) or *multivalent* (also called *polyvalent*). A monovalent vaccine is designed to immunize against a single antigen or single microorganism. A multivalent or polyvalent vaccine is designed to immunize against two or more strains of the same microorganism, or against two or more microorganisms. In certain cases a monovalent vaccine may be preferable for rapidly developing a strong immune response.

Developing Immunity

The immune system recognises vaccine agents as foreign, destroys them, and "remembers" them. When the virulent version of an agent comes along the body recognises the protein coat on the virus, and thus is prepared to respond, by (1) neutralizing the target agent before it can enter cells, and (2) by recognising and destroying infected cells before that agent can multiply to vast numbers. When two or more vaccines are mixed together in the same formulation, the two vaccines can interfere.

This most frequently occurs with live attenuated vaccines, where one of the vaccine components is more robust than the others and suppresses the growth and immune response to the other components. This phenomenon was first noted in the trivalent Sabin polio vaccine, where the amount of serotype 2 virus in the vaccine had to be reduced to stop it from interfering with the "take" of the serotype 1 and 3 viruses in the vaccine. This phenomenon has also been found to be a problem with the dengue vaccines currently being researched, where the DEN-3 serotype was found to predominate and suppress the response to DEN-1, -2 and -4 serotypes.

Vaccines have contributed to the eradication of smallpox, one of the most contagious and deadly diseases known to man. Other diseases such as rubella, polio, measles, mumps, chickenpox, and typhoid are nowhere near as common as they were a hundred years ago. As long as the vast majority of people are vaccinated, it is much more difficult for an outbreak of disease to occur, let alone spread.

This effect is called herd immunity. Polio, which is transmitted only between humans, is targeted by an extensive eradication campaign that has seen endemic polio restricted to only parts of four countries (Afghanistan, India, Nigeria and Pakistan). The difficulty of reaching all children as well as cultural misunderstandings, however, have caused the anticipated eradication date to be missed several times.

Schedule

In order to provide best protection, children are recommended to receive vaccinations as soon as their immune systems are sufficiently developed to respond to particular vaccines, with additional "booster" shots often required to achieve "full immunity". This has led to the development of complex vaccination schedules.

In the United States, the Advisory Committee on Immunization Practices, which recommends schedule additions for the Centres for Disease Control and Prevention, recommends routine vaccination of children against: hepatitis A, hepatitis B, polio, mumps, measles, rubella, diphtheria, pertussis, tetanus, HiB, chickenpox, rotavirus, influenza, meningococcal disease and pneumonia.

The large number of vaccines and boosters recommended (up to 24 injections by age two) has led to problems with achieving full compliance. In order to combat declining compliance rates, various notification systems have been instituted and a number of combination injections are now marketed (e.g., Pneumococcal conjugate vaccine and MMRV vaccine), which provide protection against multiple diseases.

Besides recommendations for infant vaccinations and boosters, many specific vaccines are recommended at other ages or for repeated injections throughout life—most commonly for measles, tetanus, influenza, and pneumonia. Pregnant women are often screened for continued resistance to rubella. The human papillomavirus vaccine is recommended in the U.S. (as of 2011) and UK (as of 2009). Vaccine recommendations for the elderly concentrate on pneumonia and influenza, which are more deadly to that group. In 2006, a vaccine was introduced against shingles, a disease caused by the chickenpox virus, which usually affects the elderly.

Effectiveness

Vaccines do not guarantee complete protection from a disease. Sometimes, this is because the host's immune system simply does not respond adequately or at all. This may be due to a lowered immunity in general (diabetes, steroid use, HIV infection, age) or because the host's immune system does not have a B cell capable of generating antibodies to that antigen.

Even if the host develops antibodies, the human immune system is not perfect and in any case the immune system might still not be able to defeat the infection immediately. In this case, the infection will be less severe and heal faster.

Adjuvants are typically used to boost immune response. Most often aluminium adjuvants are used, but adjuvants like squalene are also used in some vaccines and more vaccines with squalene and phosphate adjuvants are being tested. Larger doses are used in some cases for older people (50–75 years and up), whose immune response to a given vaccine is not as strong. The efficacy or performance of the vaccine is dependent on a number of factors:

- the disease itself (for some diseases vaccination performs better than for other diseases)
- the strain of vaccine (some vaccinations are for different strains of the disease)
- whether one kept to the timetable for the vaccinations
- some individuals are "non-responders" to certain vaccines, meaning that they do not generate antibodies even after being vaccinated correctly
- other factors such as ethnicity, age, or genetic predisposition.

When a vaccinated individual does develop the disease vaccinated against, the disease is likely to be milder than without vaccination.

The following are important considerations in the effectiveness of a vaccination program:

1. careful modelling to anticipate the impact that an immunization campaign will have on the epidemiology of the disease in the medium to long term
2. ongoing surveillance for the relevant disease following introduction of a new vaccine and
3. maintaining high immunization rates, even when a disease has become rare.

In 1958 there were 763,094 cases of measles and 552 deaths in the United States. With the help of new vaccines, the number of cases dropped to fewer than 150 per year (median of 56). In early 2008, there were 64 suspected cases of measles. 54 out of 64 infections were associated with importation from another country, although only 13% were actually acquired outside of the United States; 63 of these 64 individuals either had never been vaccinated against measles, or were uncertain whether they had been vaccinated.

Opposition to Vaccination

Opposition to vaccination, from a wide array of vaccine critics, has existed since the earliest vaccination campaigns. Although the benefits of preventing suffering and death from serious infectious diseases greatly outweigh the risks of rare adverse effects following immunization, disputes have arisen over the morality, ethics, effectiveness, and safety of vaccination. Some vaccination critics say that vaccines are ineffective against disease or that vaccine safety studies are inadequate. Some religious groups do not allow vaccination, and some political groups oppose mandatory vaccination on the grounds of individual liberty. In response, concern has been raised that spreading unfounded information about the medical risks of vaccines increases rates of life-threatening infections, not only in the children whose parents refused vaccinations, but also in other children, perhaps too young for vaccines, who could contract infections from unvaccinated carriers.

Economics of Development

One challenge in vaccine development is economic: many of the diseases most demanding a vaccine, including HIV, malaria and tuberculosis, exist principally in poor countries. Pharmaceutical firms and biotechnology companies have little incentive to develop vaccines for these diseases, because there is little revenue potential. Even in more affluent countries, financial returns are usually minimal and the financial and other risks are great.

Most vaccine development to date has relied on "push" funding by government, universities and non-profit organisations. Many vaccines have been highly cost effective and beneficial for public health. The number of vaccines actually administered has risen dramatically in recent decades.

This increase, particularly in the number of different vaccines administered to children before entry into schools may be due to government mandates and support, rather than economic incentive.

Patents

The filing of patents on vaccine development processes can also be viewed as an obstacle to the development of new vaccines. Because of the weak protection offered through a patent on the final product, the protection of the innovation regarding vaccines is often made through the patent of processes used on the development of new vaccines as well as the protection of secrecy.

Production

Vaccine production has several stages. First, the antigen itself is generated. Viruses are grown either on primary cells such as chicken eggs (*e.g.*, for influenza), or on continuous cell lines such as cultured human cells (*e.g.*, for hepatitis A). Bacteria are grown in bioreactors (*e.g.*, *Haemophilus influenzae* type b). Alternatively, a recombinant protein derived from the viruses or bacteria can be generated in yeast, bacteria, or cell cultures.

After the antigen is generated, it is isolated from the cells used to generate it. A virus may need to be inactivated, possibly with no further purification required. Recombinant proteins need many operations involving ultrafiltration and column chromatography. Finally, the vaccine is formulated by adding adjuvant, stabilizers, and preservatives as needed.

The adjuvant enhances the immune response of the antigen, stabilizers increase the storage life, and preservatives allow the use of multidose vials. Combination vaccines are harder to develop and produce, because of potential incompatibilities and interactions among the antigens and other ingredients involved. Vaccine production techniques are evolving.

Cultured mammalian cells are expected to become increasingly important, compared to conventional options such as chicken eggs, due to greater productivitity and low incidence of problems with contamination. Recombination technology that produces genetically detoxified vaccine is expected to grow in popularity for the production of bacterial vaccines that use toxoids.

Combination vaccines are expected to reduce the quantities of antigens they contain, and thereby decrease undesirable interactions, by using pathogen-associated molecular patterns.

Excipients

Beside the active vaccine itself, the following excipients are commonly present in vaccine preparations:

- Aluminium salts or gels are added as adjuvants. Adjuvants are added to promote an earlier, more potent response, and more persistent immune response to the vaccine; they allow for a lower vaccine dosage.
- Antibiotics are added to some vaccines to prevent the growth of bacteria during production and storage of the vaccine.
- Egg protein is present in influenza and yellow fever vaccines as they are prepared using chicken eggs. Other proteins may be present.
- Formaldehyde is used to inactivate bacterial products for toxoid vaccines. Formaldehyde is also used to kill unwanted viruses and bacteria that might contaminate the vaccine during production.
- Monosodium glutamate (MSG) and 2-phenoxyethanol are used as stabilizers in a few vaccines to help the vaccine remain unchanged when the vaccine is exposed to heat, light, acidity, or humidity.
- Thimerosal is a mercury-containing preservative that is added to vials of vaccine that contain more than one dose to prevent contamination and growth of potentially harmful bacteria.

Role of Preservatives

Many vaccines need preservatives to prevent serious adverse effects such as *Staphylococcus* infection that, in one 1928 incident, killed 12 of 21 children inoculated with a diphtheria vaccine that lacked a preservative.

Several preservatives are available, including thiomersal, phenoxyethanol, and formaldehyde. Thiomersal is more effective against bacteria, has better shelf life, and improves vaccine stability, potency, and safety, but in the U.S., the European Union, and a few other affluent countries, it is no longer used as a preservative in childhood vaccines, as a precautionary measure due to its mercury content. Although controversial claims have been made that thiomersal contributes to autism, no convincing scientific evidence supports these claims.

Delivery Systems

There are several new delivery systems in development that will hopefully make vaccines more efficient to deliver. Possible methods include liposomes and *ISCOM* (immune stimulating complex).

The latest developments in vaccine delivery technologies have resulted in oral vaccines. A polio vaccine was developed and tested by volunteer vaccinations with no formal training; the results were very positive in that the ease of the vaccines increased dramatically. With an oral vaccine, there is no risk of blood contamination. Oral vaccines are likely to be solid which have proven to be more stable and less likely to freeze; this stability reduces the need for a "cold chain": the resources required to keep vaccines within a restricted temperature range from the manufacturing stage to the point of administration, which, in turn, will decrease costs of vaccines. Finally, a microneedle approach, which is still in stages of development, seems to be the vaccine of the future, the microneedle, which is "pointed projections fabricated into arrays that can create vaccine delivery pathways through the skin".

Plasmids

The use of plasmids has been validated in preclinical studies as a protective vaccine strategy for cancer and infectious diseases. However, in human studies this approach has failed to provide clinically relevant benefit. The overall efficacy of plasmid DNA immunization depends on increasing the plasmid's immunogenicity while also correcting for factors involved in the specific activation of immune effector cells.

Use in Veterinary Medicine

Vaccinations of animals are used both to prevent their contracting diseases and to prevent transmission of disease to humans. Both animals kept as pets and animals raised as livestock are routinely vaccinated.

In some instances, wild populations may be vaccinated. This is sometimes accomplished with vaccine-laced food spread in a disease-prone area and has been used to attempt to control rabies in raccoons. Where rabies occurs, rabies vaccination of dogs may be required by law. Other canine vaccines include canine distemper, canine parvovirus, infectious canine hepatitis, adenovirus-2, leptospirosis, bordatella, canine parainfluenza virus, and Lyme disease among others.

Trends

Vaccine development has several trends:

- Until recently, most vaccines were aimed at infants and children, but adolescents and adults are increasingly being targeted.
- Combinations of vaccines are becoming more common; vaccines containing five or more components are used in many parts of the world.
- New methods of administering vaccines are being developed, such as skin patches, aerosols via inhalation devices, and eating genetically engineered plants.
- Vaccines are being designed to stimulate innate immune responses, as well as adaptive.

- Attempts are being made to develop vaccines to help cure chronic infections, as opposed to preventing disease.
- Vaccines are being developed to defend against bioterrorist attacks such as anthrax, plague, and smallpox.
- Appreciation for sex and pregnancy differences in vaccine responses "might change the strategies used by public health officials".

Principles that govern the immune response can now be used in tailor-made vaccines against many noninfectious human diseases, such as cancers and autoimmune disorders. For example, the experimental vaccine CYT006-AngQb has been investigated as a possible treatment for high blood pressure. Factors that have impact on the trends of vaccine development include progress in translatory medicine, demographics, regulatory science, political, cultural, and social responses.

6

Gene Technology

Genetic Code

Genes generally express their functional effect through the production of proteins, which are complex molecules responsible for most functions in the cell. Proteins are made up of one or more polypeptide chains, each of which is composed of a sequence of amino acids, and the DNA sequence of a gene (through an RNA intermediate) is used to produce a specific amino acid sequence.

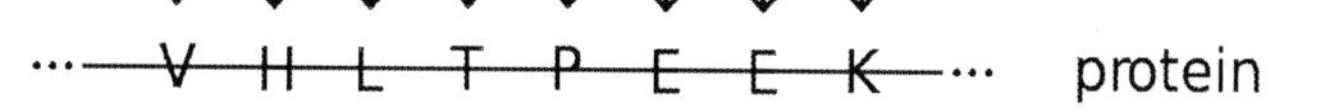

Figure: *The genetic code: DNA, through a messenger RNA intermediate, codes for protein with a triplet code.*

This process begins with the production of an RNA molecule with a sequence matching the gene's DNA sequence, a process called transcription. This messenger RNA molecule is then used to produce a corresponding amino acid sequence through a process called translation. Each group of three nucleotides in the sequence, called a codon, corresponds either to one of the twenty possible amino acids in a protein or an instruction to end the amino acic sequence; this correspondence is called the genetic code. The flow of information is unidirectional: information is transferred from nucleotide sequences into the amino acid sequence of proteins, but it never transfers from protein back into the sequence of DNA—a phenomenon Francis Crick called the central dogma of molecular biology.

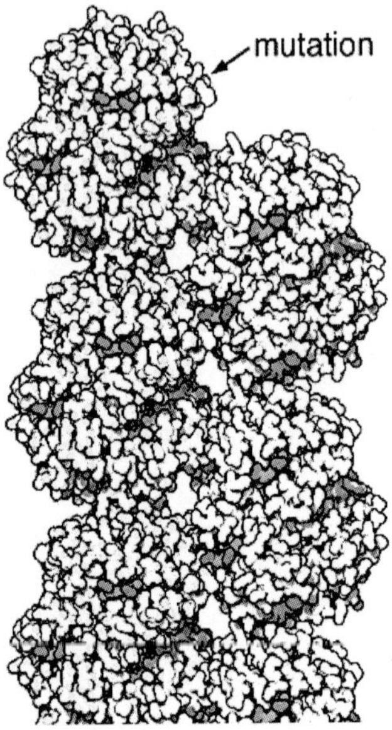

Figure: *A single amino acid change causes hemoglobin to form fibers.*

The specific sequence of amino acids results in a unique three-dimensional structure for that protein, and the three-dimensional structures of proteins are related to their functions. Some are simple structural molecules, like the fibers formed by the protein collagen. Proteins can bind to other proteins and simple molecules, sometimes acting as enzymes by facilitating chemical reactions within the bound molecules (without changing the structure of the protein itself). Protein structure is dynamic; the protein hemoglobin bends into slightly different forms as it facilitates the capture, transport, and release of oxygen molecules within mammalian blood.

A single nucleotide difference within DNA can cause a change in the amino acid sequence of a protein. Because protein structures are the result of their amino acid sequences, some changes can dramatically change the properties of a protein by destabilizing the structure or changing the surface of the protein in a way that changes its interaction with other proteins and molecules. For example, sickle-cell anemia is a human genetic disease that results from a single base difference within the coding region for the â-globin section of hemoglobin, causing a single amino acid change that changes hemoglobin's physical properties. Sickle-cell versions of hemoglobin stick to themselves, stacking to form fibers that distort the shape of red blood cells carrying the protein. These sickle-shaped cells no longer flow smoothly through blood vessels, having a tendency to clog or degrade, causing the medical problems associated with this disease. Some genes are transcribed into RNA but are not translated into protein products—such RNA molecules are called non-coding RNA. In some cases, these products fold into structures which are involved in critical cell functions (e.g. ribosomal RNA and transfer RNA). RNA can also have regulatory effect through hybridization interactions with other RNA molecules (e.g. microRNA).

Nature Versus Nurture

Although genes contain all the information an organism uses to function, the environment plays an important role in determining the ultimate phenotype—a phenomenon often referred to as "nature vs. nurture". The phenotype of an organism depends on the interaction of genetics with the environment. One example of this is the case of temperature-sensitive mutations. Often, a single amino acid change within the sequence of a protein does not change its behavior and interactions with other molecules, but it does destabilize the structure.

In a high temperature environment, where molecules are moving more quickly and hitting each other, this results in the protein losing its structure and failing to function. In a low temperature environment, however, the protein's structure is stable and it functions normally. This type of mutation is visible in the coat coloration of Siamese cats, where a mutation in an enzyme responsible for pigment production causes it to destabilize and lose function at high temperatures. The protein remains functional in areas of skin that are colder—legs, ears, tail, and face—and so the cat has dark fur at its extremities.

Environment also plays a dramatic role in effects of the human genetic disease phenylketonuria. The mutation that causes phenylketonuria disrupts the ability of the body to break down the amino acid phenylalanine, causing a toxic build-up of an intermediate molecule that, in turn, causes severe symptoms of progressive mental retardation and seizures. If someone with the phenylketonuria mutation follows a strict diet that avoids this amino acid, however, they remain normal and healthy.

A popular method to determine how much role nature and nurture play is to study identical and fraternal twins or siblings of multiple birth. Because identical siblings come from the same zygote they are genetically the same. Fraternal siblings however are as different genetically from one another as normal siblings. By comparing how often the twin of a set has the same disorder between fraternal and identical twins, scientists can see whether there is more of a nature or nurture effect. One famous example of a multiple birth study includes the Genain quadruplets, who were identical quadruplets all diagnosed with schizophrenia.

Gene Regulation

The genome of a given organism contains thousands of genes, but not all these genes need to be active at any given moment. A gene is expressed when it is being transcribed into mRNA (and translated into protein), and there exist many cellular methods of controlling the expression of genes such that proteins are produced only when needed by the cell. Transcription factors are regulatory proteins that bind to the start of genes, either promoting or inhibiting the transcription of the gene. Within the genome of Escherichia coli bacteria, for example, there exists a series of genes necessary for the synthesis of the amino acid tryptophan. However, when tryptophan is already available to the cell, these genes for tryptophan synthesis are no longer needed.

The presence of tryptophan directly affects the activity of the genes—tryptophan molecules bind to the tryptophan repressor (a transcription factor), changing the repressor's structure such that the repressor binds to the genes. The tryptophan repressor blocks the transcription and expression of the genes, thereby creating negative feedback regulation of the tryptophan synthesis process.

Figure: *Transcription factors bind to DNA, influencing the transcription of associated genes.*

Differences in gene expression are especially clear within multicellular organisms, where cells all contain the same genome but have very different structures and behaviors due to the expression of different sets of genes. All the cells in a multicellular organism derive from a single cell, differentiating into variant cell types in response to external and intercellular signals and gradually establishing different patterns of gene expression to create different behaviors. As no single gene is responsible for the development of structures within multicellular organisms, these patterns arise from the complex interactions between many cells.

Within eukaryotes there exist structural features of chromatin that influence the transcription of genes, often in the form of modifications to DNA and chromatin that are stably inherited by daughter cells. These features are called "epigenetic" because they exist "on top" of the DNA sequence and retain inheritance from one cell generation to the next. Because of epigenetic features, different cell types grown within the same medium can retain very different properties. Although epigenetic features are generally dynamic over the course of development, some, like the phenomenon of paramutation, have multigenerational inheritance and exist as rare exceptions to the general rule of DNA as the basis for inheritance.

Genetic Change

Mutations

During the process of DNA replication, errors occasionally occur in the polymerization of the second strand. These errors, called mutations, can have an impact on the phenotype of an organism, especially if they occur within the protein coding sequence of a gene. Error rates are usually very low—1 error in every 10–100 million bases—due to the "proofreading" ability of DNA polymerases. (Without proofreading error rates are a thousandfold higher; because many viruses rely on DNA and RNA polymerases that lack proofreading ability, they experience higher mutation rates.)

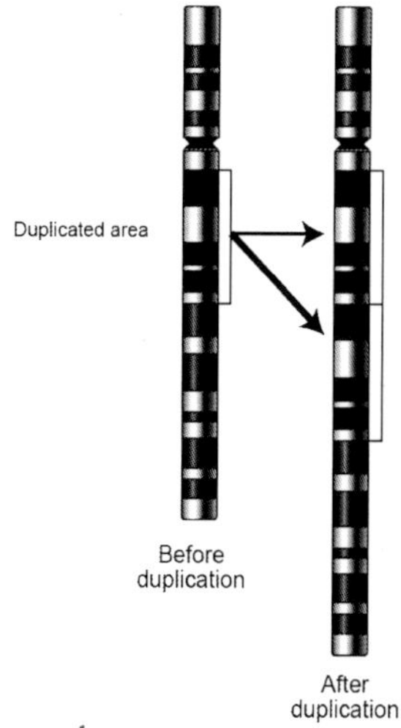

Figure: *Gene duplication allows diversification by providing redundancy: one gene can mutate and lose its original function without harming the organism.*

Processes that increase the rate of changes in DNA are called mutagenic: mutagenic chemicals promote errors in DNA replication, often by interfering with the structure of base-pairing, while UV radiation induces mutations by causing damage to the DNA structure. Chemical damage to DNA occurs naturally as well, and cells use DNA repair mechanisms to repair mismatches and breaks in DNA—nevertheless, the repair sometimes fails to return the DNA to its original sequence.

In organisms that use chromosomal crossover to exchange DNA and recombine genes, errors in alignment during meiosis can also cause mutations. Errors in crossover are especially likely when similar sequences cause partner chromosomes to adopt a mistaken alignment; this makes some regions in genomes more prone to mutating in this way. These errors create large structural changes in DNA sequence—duplications, inversions or deletions of entire regions, or the accidental exchanging of whole parts between different chromosomes (called translocation).

Natural Selection and Evolution

Natural Selection

Mutations alter an organisms genotype and occasionally this causes different phenotypes to appear. Most mutations have little effect on an organism's phenotype, health, or reproductive fitness. Mutations that do have an effect are usually deleterious, but occasionally some can be beneficial. Studies in the fly Drosophila melanogaster suggest that if a mutation changes a protein produced by a gene, about 70 percent of these mutations will be harmful with the remainder being either neutral or weakly beneficial.

Population genetics studies the distribution of genetic differences within populations and how these distributions change over time. Changes in the frequency of an allele in a population are mainly influenced by natural selection, where a given allele provides a selective or reproductive advantage to the organism, as well as other factors such as mutation, genetic drift, genetic draft, artificial selection and migration.

Over many generations, the genomes of organisms can change significantly, resulting in the phenomenon of evolution. Selection for beneficial mutations can cause a species to evolve into forms better able to survive in their environment, a process called adaptation. New species are formed through the process of speciation, often caused by geographical separations that prevent populations from exchanging genes with each other.

The application of genetic principles to the study of population biology and evolution is referred to as the modern synthesis. By comparing the homology between different species' genomes it is possible to calculate the evolutionary distance between them and when they may have diverged (called a molecular clock). Genetic comparisons are generally considered a more accurate method of characterizing the relatedness between species than the comparison of phenotypic characteristics.

The evolutionary distances between species can be used to form evolutionary trees; these trees represent the common descent and divergence of species over time, although they do not show the transfer of genetic material between unrelated species (known as horizontal gene transfer and most common in bacteria).

Research and Technology

Model Organisms

Although geneticists originally studied inheritance in a wide range of organisms, researchers began to specialize in studying the genetics of a particular subset of organisms. The fact that significant research already existed for a given organism would encourage new researchers to choose it for further study, and so eventually a few model organisms became the basis for most genetics research. Common research topics in model organism genetics include the study of gene regulation and the involvement of genes in development and cancer.

Organisms were chosen, in part, for convenience—short generation times and easy genetic manipulation made some organisms popular genetics research tools. Widely used model organisms include the gut bacterium Escherichia coli, the plant Arabidopsis thaliana, baker's yeast (Saccharomyces cerevisiae), the nematode Caenorhabditis elegans, the common fruit fly (Drosophila melanogaster), and the common house mouse (Mus musculus).

Medicine

Medical genetics seeks to understand how genetic variation relates to human health and disease. When searching for an unknown gene that may be involved in a disease, researchers commonly use genetic linkage and genetic pedigree charts to find the location on the genome associated with the disease. At the population level, researchers take advantage of Mendelian randomization to look for locations in the genome that are associated with diseases, a method especially useful for multigenic traits not clearly defined by a single gene.

Once a candidate gene is found, further research is often done on the corresponding gene (called an orthologous gene) in model organisms. In addition to studying genetic diseases, the increased availability of genotyping methods has led to the field of pharmacogenetics—studying how genotype can affect drug responses. Individuals differ in their inherited tendency to develop cancer, and cancer is a genetic disease. The process of cancer development in the body is a combination of events. Mutations occasionally occur within cells in the body as they divide. Although these mutations will not be inherited by any offspring, they can affect the behavior of cells, sometimes causing them to grow and divide more frequently. There are biological mechanisms that attempt to stop this process; signals are given to inappropriately dividing cells that should trigger cell death, but sometimes additional mutations occur that cause cells to ignore these messages. An internal process of natural selection occurs within the body and eventually mutations accumulate within cells to promote their own growth, creating a cancerous tumor that grows and invades various tissues of the body.

Normally, a cell divides only in response to signals: "growth factors", it stops growing when making contact with surrounding cells (contact inhibition), and in response to growth inhibitory signals, it divides a limited number of times and dies (apoptosis), it stays inside the epithelium and is not able to migrate to invade other organs. To become a cancer cell, a cell has to accumulate mutations in a number of genes (3-7) that allow it to bypass all these regulations: it no longer needs growth factors to divide, it continues growing when making contact to neighbor cells, and ignores inhibitory signals, it will keep growing indefinitely and is immortal, it will escape from the epithelium and ultimately may be able to escape from the primary tumor, cross the endothelium of a blood vessel, be transported by the bloodstream and will colonize a new organ, forming deadly metastasis.

Although there are some genetic predispositions in a small fraction of cancers, the major fraction is due to a set of new genetic mutations that originally appear and accumulate in one or a small number of cells that will divide to form the tumor and are not transmitted to the progeny (somatic mutations). The most frequent mutations are a loss of function of p53 protein, a tumor suppressor, or in the p53 pathway, and gain of function mutations in the ras proteins, or in other oncogenes.

For Human Genetic Diseases and Genetic Disorders

Research Methods

DNA can be manipulated in the laboratory. Restriction enzymes are commonly used enzymes that cut DNA at specific sequences, producing predictable fragments of DNA.

DNA fragments can be visualized through use of gel electrophoresis, which separates fragments according to their length. The use of ligation enzymes allows DNA fragments to be connected, and by ligating fragments of DNA together from different sources, researchers can create recombinant DNA. Often associated with genetically modified organisms, recombinant DNA is commonly used in the context of plasmids—short circular DNA fragments with a few genes on them. By inserting plasmids into bacteria and growing those bacteria on plates of agar (to isolate clones of bacteria cells), researchers can clonally amplify the inserted fragment of DNA (a process known as molecular cloning). (Cloning can also refer to creating clonal organisms, by various means.)

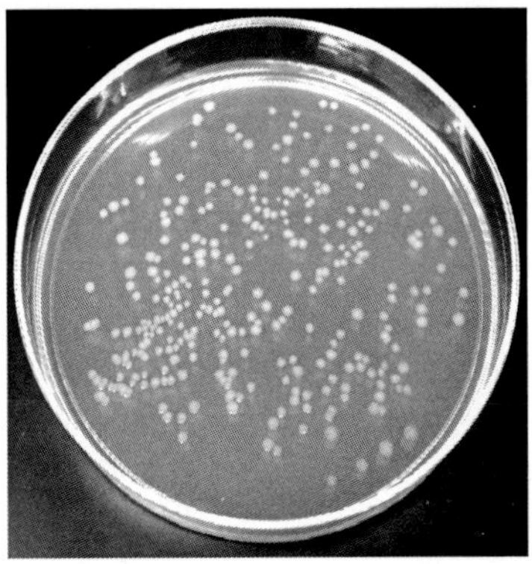

Figure: *Colonies of E. coli on a plate of agar, an example of cellular cloning and often used in molecular cloning.*

DNA can also be amplified using a procedure called the polymerase chain reaction (PCR). By using specific short sequences of DNA, PCR can isolate and exponentially amplify a targeted region of DNA. Because it can amplify from extremely small amounts of DNA, PCR is also often used to detect the presence of specific DNA sequences.

DNA Sequencing and Genomics

One of the most fundamental technologies developed to study genetics, DNA sequencing allows researchers to determine the sequence of nucleotides in DNA fragments. Developed in 1977 by Frederick Sanger and coworkers, chain-termination sequencing is now routinely used to sequence DNA fragments. With this technology, researchers have been able to study the molecular sequences associated with many human diseases.

As sequencing has become less expensive, researchers have sequenced the genomes of many organisms, using computational tools to stitch together the sequences of many different fragments (a process called genome assembly).

These technologies were used to sequence the human genome, leading to the completion of the Human Genome Project in 2003. New high-throughput sequencing technologies are dramatically lowering the cost of DNA sequencing, with many researchers hoping to bring the cost of resequencing a human genome down to a thousand dollars.

The large amount of sequence data available has created the field of genomics, research that uses computational tools to search for and analyze patterns in the full genomes of organisms. Genomics can also be considered a subfield of bioinformatics, which uses computational approaches to analyze large sets of biological data.

Quantitative Genetics

Quantitative genetics is the study of continuous traits (such as height or weight) and their underlying mechanisms. It is effectively an extension of simple Mendelian inheritance in that the combined effects of one or more genes and the environments in which they are expressed give rise to continuous distributions of phenotypic values.

History

The field was founded, in evolutionary terms, by the originators of the modern synthesis, R.A. Fisher, Sewall Wright and J. B. S. Haldane, and aimed to predict the response to selection given data on the phenotype and relationships of individuals.

Analysis of Quantitative Trait Loci, or QTL, is a more recent addition to the study of quantitative genetics. A QTL is a region in the genome that affects the trait or traits of interest. Quantitative trait loci approaches require accurate phenotypic, pedigree and genotypic data from a large number of individuals.

Basic Principles

The phenotypic value (P) of an individual is the combined effect of the genotypic value (G) and the environmental deviation (E):

$$P = G + E$$

The genotypic value is the combined effect of all the genetic effects, including nuclear genes, mitochondrial genes and interactions between the genes. It is worthwhile to note that the mathematics is related to the genetics: for which the brief following revision may be useful. In disomic (diploid) organisms, a nucleus gene is represented twice in the gene-set ("genotype"), one contribution being provided by each parent during sexual reproduction. Each "gene" is located at a particular place (a "locus" - the Latin word for place; plural "loci") on corresponding chromosomes (homologues), one from each parent. Any gene may have several functional forms in the species as a whole, and each of these may lead to outwardly different "effects" (= an average result in the phenotype considered over a large sample of gene-backgrounds and environments).

These functional forms are "alleles" (or "allelomorphs", the original term). If both alleles at a gene have the same phenotypic effect (are the "same"), the gene is said to be "homozygous": if each allele at a gene is different in effect, the gene is "heterozygous". The average phenotypic outcome may also depend upon how alleles interact with their own homologous partner in the disomic genotype ("dominance"), and on how these alleles interact with those of other genes at other loci which also affect this phenotypic trait ("epistasis"). Notice that we have combined classical genetics ideas with those of statistics in this exposition. Terms such as "gene", "homologue", "allele", "homozygous" and "heterozygous" are genetical: but "effect" is statistical, and refers to the average observed over an infinity of backgrounds, both genetical and environmental. Thus, we have very sneakily defined the "genetic value" (or "genotypic" value) as the infinity mean of all the phenotypes it can ever produce in time and space! Before molecular genetics, there simply was no other way to do it! And after molecular genetics, this is still the most utilitarian way to tackle the idea of genetic value! Also notice that we have openly used fundamentals of reproductive biology behind the genetics.

The founder of Quantitative Genetics - Sir Ronald Fisher - perceived all of this when he proposed the first mathematics of this branch of Genetics [Fisher R.A. (1930). The Genetical theory of Natural Selection. Oxford Clarendon Press.]. He sought to define a single statistical summary of all the variance arising from phenotypic change during the course of genetical assortment and segregation, which he called the "genetic" variance. His residual genotypic variance (which he called simply the "residual") represented that part of assortment which did not lead to phenotypic change, although the genes themselves had in fact been subject to meiosis and syngamy, of course. [A simplified exposition of this can be found in Falconer and Mackay (1966).] These partitions subsequently became the familiar subdivisions of the "additive" (A) and "dominance" (D) variances, respectively.

These later names are utterly misleading and very unfortunate, and have led to much confusion as to what they mean genetically. [It would have been better for posterity had they been named "Assortative" (A), and "Stable" (S).] A more gene-focused partitioning was invented by Mather and Jinks in 1971, but they also were statisticians rather than geneticists, had their own rather opaque symbolism, and became somewhat overwhelmed by the Fisherian approach.

Added to all of this was the problem that Fisher's underlying reproduction model (mating system) was unrealistically simplistic: whilst it facilitated solving the equations, it didn't describe many real-life scenarios. Fortunately, Wright (in the 1950s) did provide the means to overcome complex mating systems, but its own complexity minimised its popular adoption.

Early in this millennium [Heredity (2003) 91: 85-89], a comprehensive linking together of all of these approaches reconciled their various meanings and relationships (as well as correcting an error), and suggested that new partitions reflecting real homozygosis [a] and heterozygosis [d] should replace the present "additive" [A] and "dominance" [D] subdivisions. This paper revealed, by the way, that the Additive Genetical variance consists of all the homozygote variance, plus part of the dominance variance, and a frequency-weighted covariance between homozygote and heterozygote gene effects.

The so-called Dominance variance contains only the remainder of the overall dominance variance of the gene in question, being therefore very misleading indeed! It should be understood, however, that either method of partitioning still accounts for all of the genotypic variance in the model being used: it's the way it has been divided up which is being debated. At least the Environmental variance is much more straight-forward. This can be subdivided into a pure environmental component (E) and an interaction component (I) describing the gene-environment interaction. The overall "single gene" model can be written as:

$$P = a + d + E + I.$$

Expansion of the model to multiple genes (loci) is still not resolved satisfactorily, and until that is solved it is not possible to account for epistasis. The problem is being tackled currently. The contribution of those components cannot be determined in a single individual, but they can be estimated for whole populations by estimating the variances for those components, denoted as: $V_P = V_a + V_d + V_E + V_I$

The heritability of a trait is the proportion of the total (i.e. phenotypic) variance (V_P) that is explained by the total genotypic variance (V_G). This is known also as the "broad sense" heritability (H). If only Additive genetic variance (V_A) is used in the numerator, the heritability is "narrow sense" (h). Unfortunately, this is often simply called "heritability", with little reflection about its true meaning. The broadsense heritability indicates the genotypic determination of the phenotype: while the latter estimates the degree of assortative disequlibrium in the trait.

Fisher proposed that this narrow-sense heritability might be appropriate in considering the results of natural selection, focusing as it does on disequilibrium: and it has been used also for predicting the results of artificial selection. This latter usage seems to be inappropriate, however, as breeders are interested in steering attributes towards new phenotypes (that is in utilising all the gene effects), rather than simply exploiting disequilibrium.

Resemblance Between Relatives

Central in estimating the variances for the various components is the principle of relatedness. A child has a father and a mother. Consequently, the child and father share 50% of their alleles, as do the child and the mother. However, the mother and father normally do not share alleles as a result of shared ancestors. Similarly, two full siblings share also on average 50% of the alleles with each other, while half siblings share only 25% of their alleles.

This variation in relatedness can be used to estimate which proportion of the total phenotypic variance (V_p) is explained by the above-mentioned components.

Correlated Traits

Although some genes have only an effect on a single trait, many genes have an effect on various traits. Because of this, a change in a single gene will have an effect on all those traits. This is calculated using covariances, and the phenotypic covariance (Cov_p) between two traits can be partitioned in the same way as the variances described above.

The genetic correlation is calculated by dividing the covariance between the additive genetic effects of two traits by the square root of the product of the variances for the additive genetic effects of the two traits:

$$Genetic\ correlation = \frac{Cov(A_1, A_2)}{\sqrt{V_{A1} * V_{A_2}}}$$

Genetic Distance

Genetic distance refers to the genetic divergence between species or between populations within a species. It is measured by a variety of parameters. Smaller genetic distances indicate a close genetic relationship whereas large genetic distances indicate a more distant genetic relationship.

Genetic distance can be used to compare the genetic similarity between different species, such as humans and chimpanzees. Within a species genetic distance can be used to measure the divergence between different sub-species.

In its simplest form, the genetic distance between two populations is the difference in frequencies of a trait. For example the frequency of RH negative individuals is 50.4% among Basques, 41.2% in France and 41.1% in England.

Thus the genetic difference between the Basques and French is 9.2% and the genetic difference between the French and the English is 0.1% for the RH negative trait. The genetic distance of several individual traits can then be averaged to compute an overall genetic distance.

Measures of Genetic Distance

There are several measures used to indicate genetic distance. These include:

Fixation Index

A commonly used measure of genetic distance is the fixation index which varies between 0 and 1. A value of 0 indicates that two populations are genetically identical whereas a value of 1 indicates that two populations are different species.

Nei's Standard Genetic Distance

This measure assumes that genetic differences arise due to mutations and genetic drift.

Cavalli-Sforza and Edwards 1967

This measure assumes that genetic differences arise due to genetic drift only.

Reynolds, Weir, and Cockerham's 1983

This measure assumes that genetic differences arise due to genetic drift only.

A chromosome is an organized structure of DNA and protein found in cells. It is a single piece of coiled DNA containing many genes, regulatory elements and other nucleotide sequences. Chromosomes also contain DNA-bound proteins, which serve to package the DNA and control its functions.

Chromosomes vary widely between different organisms. The DNA molecule may be circular or linear, and can be composed of 100,000 to 10,000,000,000 nucleotides in a long chain. Typically, eukaryotic cells (cells with nuclei) have large linear chromosomes and prokaryotic cells (cells without defined nuclei) have smaller circular chromosomes, although there are many exceptions to this rule. Also, cells may contain more than one type of chromosome; for example, mitochondria in most eukaryotes and chloroplasts in plants have their own small chromosomes.

In eukaryotes, nuclear chromosomes are packaged by proteins into a condensed structure called chromatin. This allows the very long DNA molecules to fit into the cell nucleus. The structure of chromosomes and chromatin varies through the cell cycle. Chromosomes are the essential unit for cellular division and must be replicated, divided, and passed successfully to their daughter cells so as to ensure the genetic diversity and survival of their progeny. Chromosomes may exist as either duplicated or unduplicated. Unduplicated chromosomes are single linear strands, whereas duplicated chromosomes contain two identical copies (called chromatids) joined by a centromere.

Compaction of the duplicated chromosomes during mitosis and meiosis results in the classic four-arm structure (pictured to the right). Chromosomal recombination plays a vital role in genetic diversity. If these structures are manipulated incorrectly, through processes known as chromosomal instability and translocation, the cell may undergo mitotic catastrophe and die, or it may unexpectedly evade apoptosis leading to the progression of cancer.

In practice "chromosome" is a rather loosely defined term. In prokaryotes and viruses, the term genophore is more appropriate when no chromatin is present. However, a large body of work uses the term chromosome regardless of chromatin content. In prokaryotes, DNA is usually arranged as a circle, which is tightly coiled in on itself, sometimes accompanied by one or more smaller, circular DNA molecules called plasmids.

These small circular genomes are also found in mitochondria and chloroplasts, reflecting their bacterial origins. The simplest genophores are found in viruses: these DNA or RNA molecules are short linear or circular genophores that often lack structural proteins. The word chromosome comes from the Greek ÷ñõìá (chroma, colour) and óõìá (soma, body) due to their property of being very strongly stained by particular dyes.

History

In a series of experiments beginning in the mid-1880s, Theodor Boveri gave the definitive demonstration that chromosomes are the vectors of heredity. His two principles were the continuity of chromosomes and the individuality of chromosomes. It is the second of these principles that was so original. Wilhelm Roux suggested that each chromosome carries a different genetic load.

Boveri was able to test and confirm this hypothesis. Aided by the rediscovery at the start of the 1900s of Gregor Mendel's earlier work, Boveri was able to point out the connection between the rules of inheritance and the behaviour of the chromosomes.

Boveri influenced two generations of American cytologists: Edmund Beecher Wilson, Walter Sutton and Theophilus Painter were all influenced by Boveri (Wilson and Painter actually worked with him).

In his famous textbook The Cell in Development and Heredity, Wilson linked together the independent work of Boveri and Sutton (both around 1902) by naming the chromosome theory of inheritance the "Sutton-Boveri Theory" (the names are sometimes reversed). Ernst Mayr remarks that the theory was hotly contested by some famous geneticists: William Bateson, Wilhelm Johannsen, Richard Goldschmidt and T.H. Morgan, all of a rather dogmatic turn of mind. Eventually, complete proof came from chromosome maps in Morgan's own lab.

In Eukaryotes

Eukaryotes (cells with nuclei such as those found in plants, yeast, and animals) possess multiple large linear chromosomes contained in the cell's nucleus. Each chromosome has one centromere, with one or two arms projecting from the centromere, although, under most circumstances, these arms are not visible as such. In addition, most eukaryotes have a small circular mitochondrial genome, and some eukaryotes may have additional small circular or linear cytoplasmic chromosomes. In the nuclear chromosomes of eukaryotes, the uncondensed DNA exists in a semi-ordered structure, where it is wrapped around histones (structural proteins), forming a composite material called chromatin.

Chromatin

Chromatin is the complex of DNA and protein found in the eukaryotic nucleus, which packages chromosomes. The structure of chromatin varies significantly between different stages of the cell cycle, according to the requirements of the DNA.

Interphase Chromatin

During interphase (the period of the cell cycle where the cell is not dividing), two types of chromatin can be distinguished:

- Euchromatin, which consists of DNA that is active, e.g., being expressed as protein.
- Heterochromatin, which consists of mostly inactive DNA. It seems to serve structural purposes during the chromosomal stages. Heterochromatin can be further distinguished into two types:
    - o Constitutive heterochromatin, which is never expressed. It is located around the centromere and usually contains repetitive sequences.
    - o Facultative heterochromatin, which is sometimes expressed.

Individual chromosomes cannot be distinguished at this stage – they appear in the nucleus as a homogeneous tangled mix of DNA and protein.

Metaphase Chromatin and Division

In the early stages of mitosis or meiosis (cell division), the chromatin strands become more and more condensed. They cease to function as accessible genetic material (transcription stops) and become a compact transportable form.

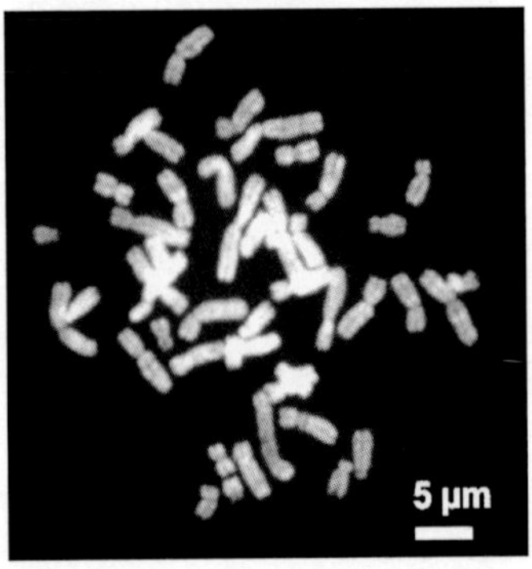

Figure: *Human chromosomes during metaphase.*

This compact form makes the individual chromosomes visible, and they form the classic four arm structure, a pair of sister chromatids attached to each other at the centromere. The shorter arms are called p arms (from the French petit, small) and the longer arms are called q arms (q follows p in the Latin alphabet; q-g "grande"). This is the only natural context in which individual chromosomes are visible with an optical microscope.

During mitosis, microtubules grow from centrosomes located at opposite ends of the cell and also attach to the centromere at specialized structures called kinetochores, one of which is present on each sister chromatid. A special DNA base sequence in the region of the kinetochores provides, along with special proteins, longer-lasting attachment in this region.

The microtubules then pull the chromatids apart toward the centrosomes, so that each daughter cell inherits one set of chromatids. Once the cells have divided, the chromatids are uncoiled and DNA can again be transcribed.

In spite of their appearance, chromosomes are structurally highly condensed, which enables these giant DNA structures to be contained within a cell nucleus.

In Humans

Chromosomes can be divided into two types—autosomes, and sex chromosomes. Certain genetic traits are linked to a person's sex and are passed on through the sex chromosomes.

The autosomes contain the rest of the genetic hereditary information. All act in the same way during cell division. Human cells have 23 pairs of large linear nuclear chromosomes (22 pairs of autosomes and one pair of sex chromosomes), giving a total of 46 per cell. In addition to these, human cells have many hundreds of copies of the mitochondrial genome.

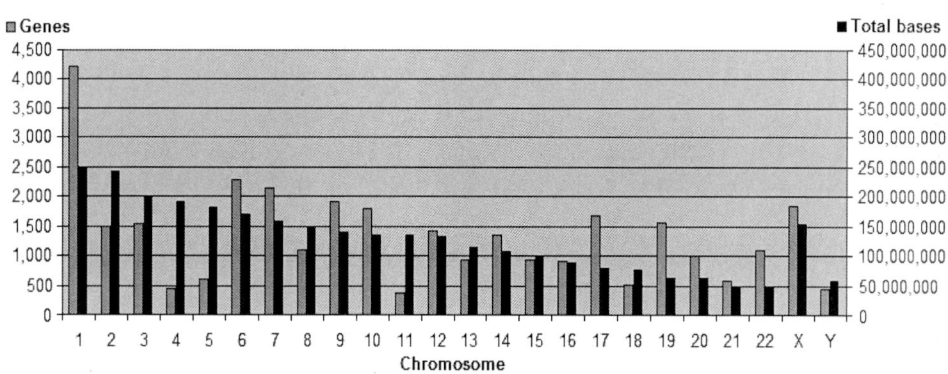

Figure: *Estimated number of genes and base pairs on each human chromosome. The right-hand legend of this chart is mislabeled "Total bases". It should say "Total base pairs".*

Sequencing of the human genome has provided a great deal of information about each of the chromosomes. Below is a table compiling statistics for the chromosomes, based on the Sanger Institute's human genome information in the Vertebrate Genome Annotation (VEGA) database.

Number of genes is an estimate as it is in part based on gene predictions. Total chromosome length is an estimate as well, based on the estimated size of unsequenced heterochromatin regions.

In Prokaryotes

The prokaryotes – bacteria and archaea – typically have a single circular chromosome, but many variations exist. Most bacteria have a single circular chromosome that can range in size from only 160,000 base pairs in the endosymbiotic bacterium Candidatus Carsonella ruddii, to 12,200,000 base pairs in the soil-dwelling bacterium Sorangium cellulosum.

Spirochaetes of the genus Borrelia are a notable exception to this arrangement, with bacteria such as Borrelia burgdorferi, the cause of Lyme disease, containing a single linear chromosome.

Structure in Sequences

Prokaryotic chromosomes have less sequence-based structure than eukaryotes. Bacteria typically have a single point (the origin of replication) from which replication starts, whereas some archaea contain multiple replication origins. The genes in prokaryotes are often organized in operons, and do not usually contain introns, unlike eukaryotes.

DNA Packaging

Prokaryotes do not possess nuclei. Instead, their DNA is organized into a structure called the nucleoid.

The nucleoid is a distinct structure and occupies a defined region of the bacterial cell. This structure is, however, dynamic and is maintained and remodeled by the actions of a range of histone-like proteins, which associate with the bacterial chromosome. In archaea, the DNA in chromosomes is even more organized, with the DNA packaged within structures similar to eukaryotic nucleosomes.

Bacterial chromosomes tend to be tethered to the plasma membrane of the bacteria. In molecular biology application, this allows for its isolation from plasmid DNA by centrifugation of lysed bacteria and pelleting of the membranes (and the attached DNA).

Prokaryotic chromosomes and plasmids are, like eukaryotic DNA, generally supercoiled. The DNA must first be released into its relaxed state for access for transcription, regulation, and replication.

Number of Chromosomes in Various Organisms

Eukaryotes

These tables give the total number of chromosomes (including sex chromosomes) in a cell nucleus. For example, human cells are diploid and have 22 different types of autosome, each present as two copies, and two sex chromosomes.

This gives 46 chromosomes in total. Other organisms have more than two copies of their chromosomes, such as bread wheat, which is hexaploid and has six copies of seven different chromosomes – 42 chromosomes in total.

Normal members of a particular eukaryotic species all have the same number of nuclear chromosomes. Other eukaryotic chromosomes, i.e., mitochondrial and plasmid-like small chromosomes, are much more variable in number, and there may be thousands of copies per cell.

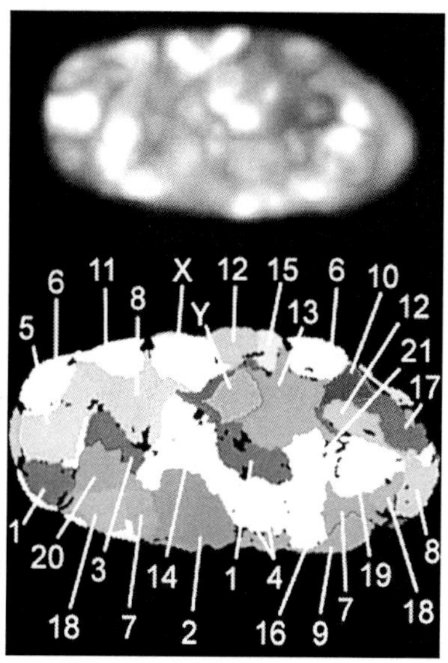

Figure: *The 23 human chromosome territories during prometaphase in fibroblast cells.*

Asexually reproducing species have one set of chromosomes, which are the same in all body cells. However, asexual species can be either haploid or diploid.

Sexually reproducing species have somatic cells (body cells), which are diploid [2n] having two sets of chromosomes, one from the mother and one from the father.

Gametes, reproductive cells, are haploid [n]: They have one set of chromosomes. Gametes are produced by meiosis of a diploid germ line cell. During meiosis, the matching chromosomes of father and mother can exchange small parts of themselves (crossover), and thus create new chromosomes that are not inherited solely from either parent. When a male and a female gamete merge (fertilization), a new diploid organism is formed. Some animal and plant species are polyploid [Xn]: They have more than two sets of homologous chromosomes. Plants important in agriculture such as tobacco or wheat are often polyploid, compared to their ancestral species.

Wheat has a haploid number of seven chromosomes, still seen in some cultivars as well as the wild progenitors. The more-common pasta and bread wheats are polyploid, having 28 (tetraploid) and 42 (hexaploid) chromosomes, compared to the 14 (diploid) chromosomes in the wild wheat.

Prokaryotes

Prokaryote species generally have one copy of each major chromosome, but most cells can easily survive with multiple copies. For example, Buchnera, a symbiont of aphids has multiple copies of its chromosome, ranging from 10–400 copies per cell.

However, in some large bacteria, such as Epulopiscium fishelsoni up to 100,000 copies of the chromosome can be present. Plasmids and plasmid-like small chromosomes are, as in eukaryotes, very variable in copy number. The number of plasmids in the cell is almost entirely determined by the rate of division of the plasmid – fast division causes high copy number.

Karyotype

In general, the karyotype is the characteristic chromosome complement of a eukaryote species. The preparation and study of karyotypes is part of cytogenetics.

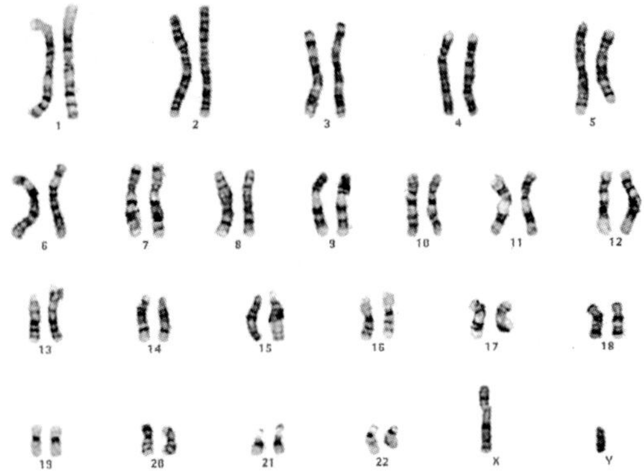

Figure: *Karyogram of a human male*

Although the replication and transcription of DNA is highly standardized in eukaryotes, the same cannot be said for their karyotypes, which are often highly variable. There may be variation between species in chromosome number and in detailed organization. In some cases, there is significant variation within species. Often there is:

1. variation between the two sexes
2. variation between the germ-line and soma (between gametes and the rest of the body)
3. variation between members of a population, due to balanced genetic polymorphism
4. geographical variation between races
5. mosaics or otherwise abnormal individuals.

Also, variation in karyotype may occur during development from the fertilised egg. The technique of determining the karyotype is usually called karyotyping. Cells can be locked part-way through division (in metaphase) in vitro (in a reaction vial) with colchicine. These cells are then stained, photographed, and arranged into a karyogram, with the set of chromosomes arranged, autosomes in order of length, and sex chromosomes (here X/Y) at the end:

Like many sexually reproducing species, humans have special gonosomes (sex chromosomes, in contrast to autosomes). These are XX in females and XY in males.

Historical Note

Investigation into the human karyotype took many years to settle the most basic question: How many chromosomes does a normal diploid human cell contain? In 1912, Hans von Winiwarter reported 47 chromosomes in spermatogonia and 48 in oogonia, concluding an XX/XO sex determination mechanism. Painter in 1922 was not certain whether the diploid number of man is 46 or 48, at first favouring 46.

He revised his opinion later from 46 to 48, and he correctly insisted on humans having an XX/XY system.

New techniques were needed to definitively solve the problem:

1. Using cells in culture
2. Pretreating cells in a hypotonic solution, which swells them and spreads the chromosomes
3. Arresting mitosis in metaphase by a solution of colchicine
4. Squashing the preparation on the slide forcing the chromosomes into a single plane
5. Cutting up a photomicrograph and arranging the result into an indisputable karyogram.

It took until 1954 before the human diploid number was confirmed as 46. Considering the techniques of Winiwarter and Painter, their results were quite remarkable. Chimpanzees (the closest living relatives to modern humans) have 48 chromosomes.

Aberrations

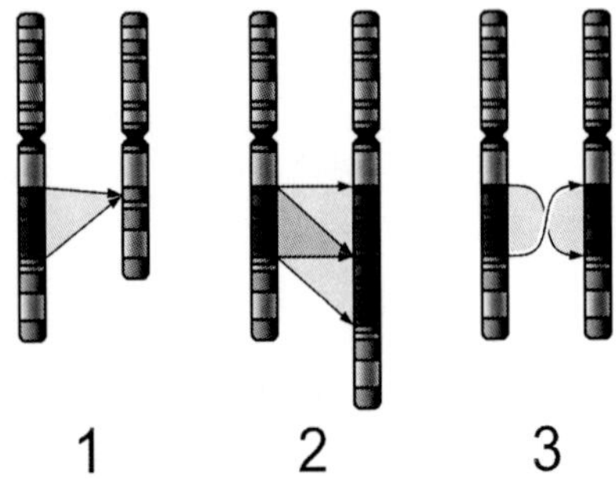

Figure: *The three major single chromosome mutations; deletion (1), duplication (2) and inversion (3).*

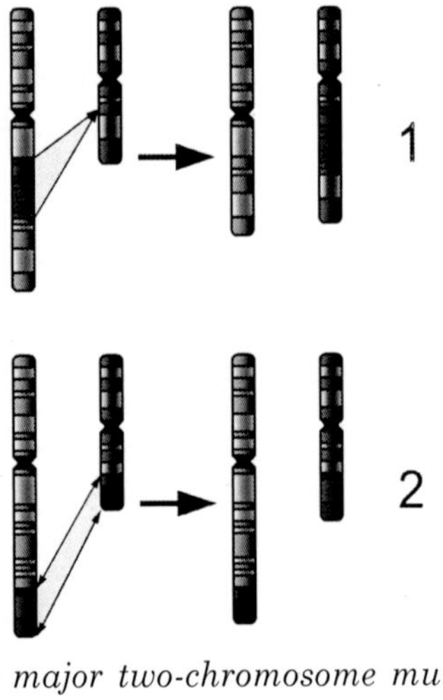

Figure: *The two major two-chromosome mutations; insertion (1) and translocation (2).*

In Down Syndrome, there are three Copies of Chromosome 21

Chromosomal aberrations are disruptions in the normal chromosomal content of a cell and are a major cause of genetic conditions in humans, such as Down syndrome, although most aberrations have little to no effect. Some chromosome abnormalities do not cause disease in carriers, such as translocations, or chromosomal inversions, although they may lead to a higher chance of birthing a child with a chromosome disorder. Abnormal numbers of chromosomes or chromosome sets, called aneuploidy, may be lethal or may give rise to genetic disorders. Genetic counseling is offered for families that may carry a chromosome rearrangement. The gain or loss of DNA from chromosomes can lead to a variety of genetic disorders. Human examples include:

- Cri du chat, which is caused by the deletion of part of the short arm of chromosome 5. "Cri du chat" means "cry of the cat" in French; the condition was so-named because affected babies make high-pitched cries that sound like those of a cat. Affected individuals have wide-set eyes, a small head and jaw, moderate to severe mental health issues, and are very short.
- Down syndrome, the most common trisomy, usually caused by an extra copy of chromosome 21 (trisomy 21). Characteristics include decreased muscle tone, stockier build, asymmetrical skull, slanting eyes and mild to moderate developmental disability.
- Edwards syndrome, or trisomy-18, the second-most-common trisomy. Symptoms include motor retardation, developmental disability and numerous congenital anomalies causing serious health problems. Ninety percent of those affected die in infancy; however, those that live past their first birthday are usually quite healthy thereafter. They have characteristic clenched hands and overlapping fingers.
- Isodicentric 15, also called idic(15), partial tetrasomy 15q, or inverted duplication 15 (inv dup 15).
- Jacobsen syndrome, which is very rare. It is also called the terminal 11q deletion disorder. Those affected have normal intelligence or mild developmental disability, with poor expressive language skills. Most have a bleeding disorder called Paris-Trousseau syndrome.
- Klinefelter's syndrome (XXY). Men with Klinefelter syndrome are usually sterile, and tend to be taller and have longer arms and legs than their peers. Boys with the syndrome are often shy and quiet, and have a higher incidence of speech delay and dyslexia. Without testosterone treatment, some may develop gynecomastia during puberty.
- Patau Syndrome, also called D-Syndrome or trisomy-13. Symptoms are somewhat similar to those of trisomy-18, without the characteristic folded hand.
- Small supernumerary marker chromosome. This means there is an extra, abnormal chromosome. Features depend on the origin of the extra genetic material. Cat-eye syndrome and isodicentric chromosome 15 syndrome (or Idic15) are both caused by a supernumerary marker chromosome, as is Pallister-Killian syndrome.
- Triple-X syndrome (XXX). XXX girls tend to be tall and thin and have a higher incidence of dyslexia.

- Turner syndrome (X instead of XX or XY). In Turner syndrome, female sexual characteristics are present but underdeveloped. Females with Turner syndrome often have a short stature, low hairline, abnormal eye features and bone development and a "caved-in" appearance to the chest.
- XYY syndrome. XYY boys are usually taller than their siblings. Like XXY boys and XXX girls, they are more likely to have learning difficulties.
- Wolf-Hirschhorn syndrome, which is caused by partial deletion of the short arm of chromosome 4. It is characterized by severe growth retardation and severe to profound mental health issues.

7

Genetic Mechanisms and Molecular Biology

In addition to very frequent simple mutations, there are several mutation classes which involve sequence exchange between allelic or nonallelic sequences, often involving repeated sequences. For example, tandemly repetitive DNA is prone to deletion/insertion polymorphism whereby different alleles vary in the number of integral copies of the tandem repeat. Such variable number of tandem repeat (VNTR) polymorphisms can occur in the case of repeated units that are very short (microsatellites); intermediate (minisatellites) or large. Different genetic mechanisms can account for VNTR polymorphism depending on the size of the repeating unit. In addition, interspersed repeats can also predispose to deletions/duplications by a variety of different genetic mechanisms.

Slipped Strand Mispairing can Cause VNTR Polymorphism at Short Tandem Repeats (Microsatellites)

There is considerable variation in the germline mutation rates at microsatellite loci, ranging from an undetectable level up to about 8×10^{-3}. Novel length alleles at (CA)/ (TG) microsatellites and at tetranucleotide marker loci are known to be formed without exchange of flanking markers.

This means that they are not generated by unequal crossover. Instead, as new mutant alleles have been observed to differ by a single repeat unit from the originating parental allele (Mahtani and Willard, 1993), the most likely mechanism to explain length variation is a form of exchange of sequence information which commences by slipped strand mispairing.

This occurs when the normal pairing between the two complementary strands of a double helix is altered by staggering of the repeats on the two strands, leading to incorrect pairing of repeats. Although slipped strand mispairing can be envisaged to occur in nonreplicating DNA, replicating DNA may offer more opportunity for slippage and hence the mechanism is often also called replication slippage or polymerase slippage. In addition to mispairing between tandem repeats, slippage replication has been envisaged to generate large deletions and duplications by mispairing between noncontiguous repeats and has been suggested to be a major mechanism for DNA sequence and genome evolution. The pathogenic potential of short tandem repeats is considerable.

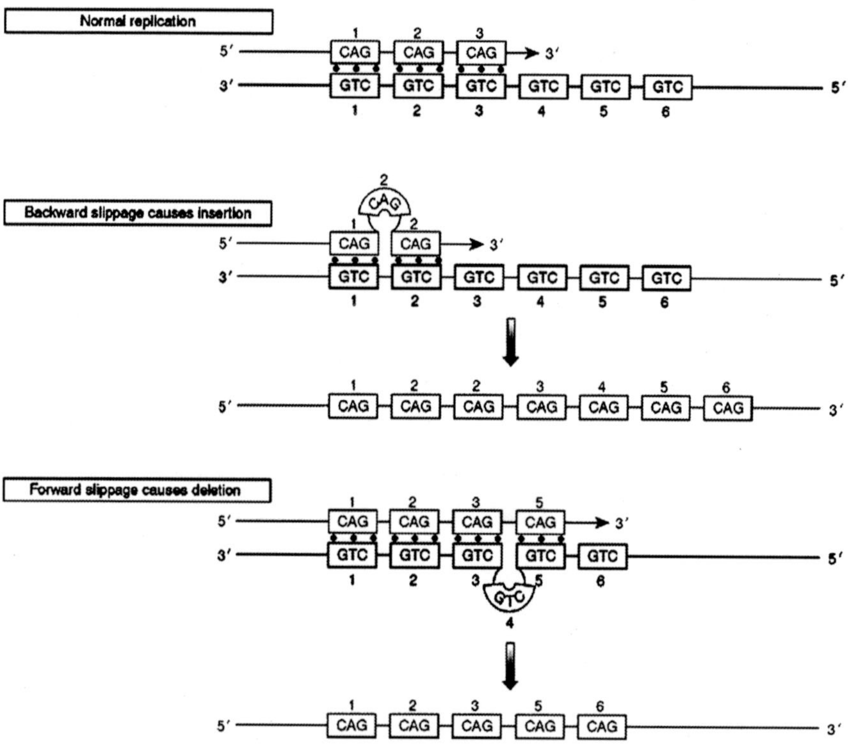

Figure: *Slipped strand mispairing during DNA replication can cause insertions or deletions.*

Large Units of Tandemly Repeated DNA are Prone to Insertion/deletion as a Result of Unequal Crossover or Unequal Sister Chromatid Exchanges

Homologous recombination describes recombination (crossover) occurring at meiosis or, rarely, mitosis between identical or very similar DNA sequences. It usually involves breakage of *nonsister chromatids* of a pair of homologs and rejoining of the fragments to generate new recombinant strands. Sister chromatid exchange is an analogous type of sequence exchange involving breakage of individual *sister chromatids* and rejoining fragments that initially were on different chromatids of the same chromosome. Both homologous recombination and sister chromatid exchange normally involve *equal* exchanges - cleavage and rejoining of the chromatids occurs at the same position on each chromatid. As a result, the exchanges occur *between allelic sequences* and at corresponding positions within alleles. In the case of intragenic equal crossover between two alleles, a new allele can result which is a fusion gene (or hybrid gene), comprising a terminal fragment from one allele and the remaining sequence of the second allele. However, equal sister chromatid exchanges cannot normally produce genetic variation because sister chromatids have identical DNA sequences.

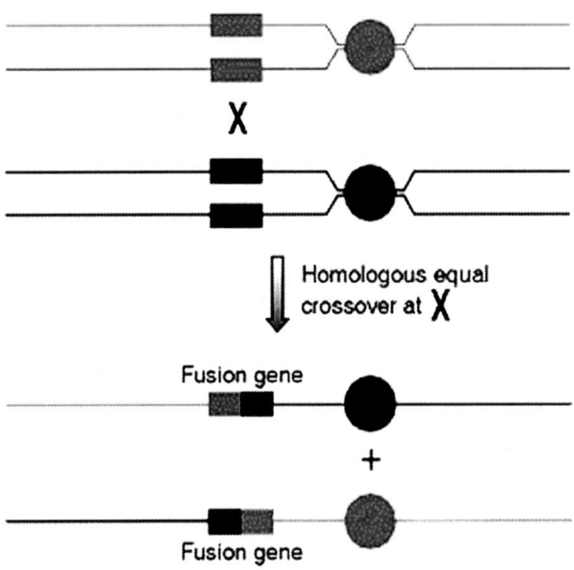

Figure: *Homologous equal crossover can result in fusion genes.*

Unequal crossover is a form of recombination in which the crossover takes place *between nonallelic sequences* on nonsister chromatids of a pair of homologs. Often the sequences at which crossover takes place show very considerable sequence homology which presumably stabilizes mispairing of the chromosomes. Because crossover occurs between mispaired nonsister chromatids, the exchange results in a deletion on one of the participating chromatids and an insertion on the other.

The analogous exchange between sister chromatids is called unequal sister chromatid exchange. Both mechanisms occur predominantly at locations where the tandemly repeated units are moderate to large in size. In such cases, the very high degree of sequence homology between the different repeats can facilitate pairing of nonallelic repeats on nonsister chromatids or sister chromatids. If chromosome breakage and rejoining occurs while the chromatids are mispaired in this way, an insertion or deletion of an integral number of repeat units will result. Note that such exchanges are *reciprocal*; both participating chromatids are modified, in one case resulting in an insertion, and in the other case in a complementary deletion.

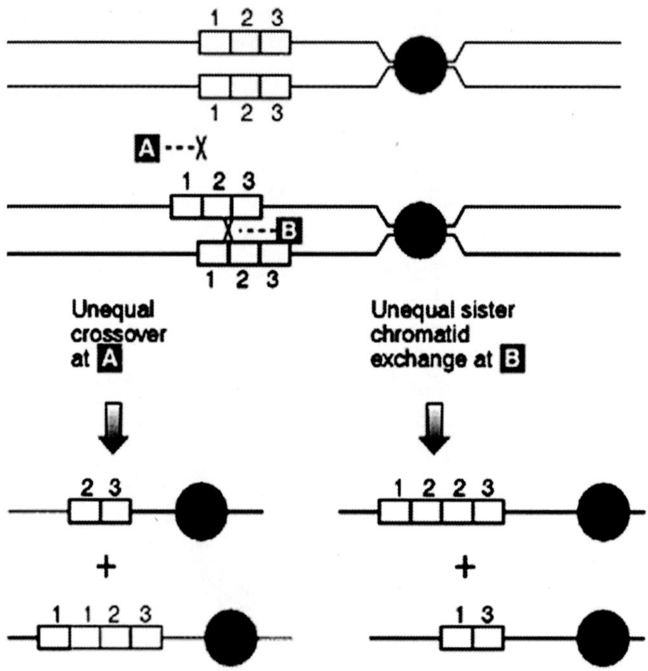

Figure: *Unequal crossover and unequal sister chromatid exchange cause insertions and deletions.*

Unequal sister chromatid exchange is thought to be a major mechanism underlying VNTR polymorphism in the rDNA clusters. Unequal crossover is also expected to occur comparatively frequently in complex satellite DNA repeats and at tandemly repeated gene loci. In the latter case, unequal crossover is known to generate pathogenic deletions at some loci. Such exchanges can also lead to concerted evolution by causing a particular variant to spread through an array of tandem repeats, resulting in *homogenisation* of the repeat units.

Occasionally, unequal crossover and unequal sister chromatid exchanges can occur at regions where there is little homology. This is likely to be the case when such mechanisms first generate a tandemly duplicated locus following mispairing of nonallelic repeats such as two *Alu* repeats or even smaller elements.

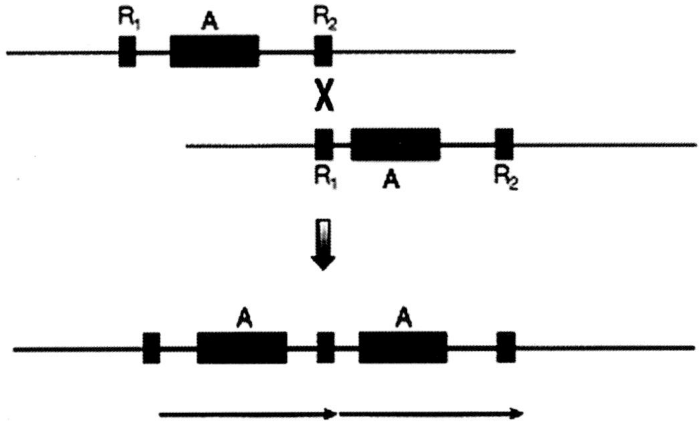

Figure: *Tandem gene duplication can result from unequal crossover or unequal sister chromatid exchange, facilitated by short interspersed repeats.*

Gene Conversion Events may be Relatively Frequent in Tandemly Repetitive DNA

Gene conversion describes a *nonreciprocal* transfer of sequence information between a pair of nonallelic DNA sequences (*interlocus gene conversion)* or allelic sequences (*interallelic gene conversion*). One of the pair of interacting sequences, the *donor*, remains unchanged. The other DNA sequence, the *acceptor*, is changed by having some or all of its sequence replaced by a sequence copied from the donor sequence. The sequence exchange is therefore a directional one; the acceptor sequence is modified by the donor sequence, but not the other way round.

One possible mechanism for gene conversion envisages formation of a heteroduplex between a DNA strand from the donor gene and a complementary strand from the acceptor gene. Following heteroduplex formation, conversion of an acceptor gene segment may occur by mismatch repair - DNA repair enzymes recognize that the two strands of the heteroduplex are not perfectly matched and 'correct' the DNA sequence of the acceptor strand to make it perfectly complementary in the converted region to the sequence of the donor gene strand.

Gene conversion has been well-described in fungi where all four products of meiosis can be recovered and studied (tetrad analysis). In humans and mammals it is not possible to do this and so gene conversion cannot be demonstrated unambiguously in higher organisms (it can never be distinguished from double crossover events, for example, although double crossovers occurring in very close proximity would normally be expected to be extremely unlikely). Despite the difficulty in identifying gene conversion in complex organisms, there are numerous instances in mammalian genomes where an allele at one locus shows a pattern of mutations which strongly resembles those found in alleles at another locus of the same species. Such evidence suggests gene conversion-like exchanges between loci.

Although simple comparisons of two sequences may be suggestive, the evidence for gene conversion is most compelling when a new mutant allele can be compared directly with its progenitor sequence. Certain highly mutable loci lend themselves to this type of analysis. In particular, some hypervariable minisatellite loci have high germline mutation rates (often 1% or more per gamete) and individual repeats often show nucleotide differences so that repeat subclasses can be recognised.

Germline mutations can be studied by detecting and characterising mutant minisatellite alleles in individual gametes. To do this, PCR analysis has been conducted on multiple dilute aliquots of DNA isolated from the sperm of an individual (small pool PCR), where each aliquot is calibrated to contain a few, perhaps 100, input molecules (Jeffreys *et al.*, 1994). The PCR products recovered from individual pools can then be typed to identify any new mutations that result in a novel allele whose length is sufficiently different as to be distinguishable from the progenitor allele.

Analyses of the patterns of germline mutation at three such loci have failed to identify exchanges of flanking markers and have shown that most mutations occurring at these loci are polar, involving the preferential gain of a few repeats at one end of a tandem repeat array. There is a bias towards gain of repeats and evidence was obtained for nonreciprocal sequence exchange between alleles, suggesting interallelic gene conversion (Jeffreys *et al.*, 1994). Evidence for interlocus gene conversion has also been obtained in human genes, notably the steroid 21-hydroxylase gene.

Pathogenic Mutations

There is a High Deleterious Mutation Rate in Hominids

Neutral mutations (those which are neither detrimental nor advantageous for the organism carrying them) accumulate throughout the generations at a rate equal to the mutation rate. To get an estimate of the (total) mutation rate is therefore simple: all one needs to do is to measure the rate of change of some presumed neutral sequence (e.g. intronic, pseudogene, etc). The deleterious mutation rate, by contrast, has been notoriously difficult to measure and no convincing estimate existed for any vertebrate until a study reported by Eyre-Walker and Keightley (1999). They investigated amino acid changes in 46 proteins occurring in the human ancestral line after its divergence from the chimpanzee. If all non-synonymous substitutions were neutral, 231 new substitutions would have been expected in their sample of genes (given an average neutral mutation rate of 0.0056 nonsynonymous substitutions per nucleotide and a total of 41 471 nucleotides investigated). Instead, only 143 nonsynonymous substitutions were observed and 88 such substitutions were inferred to have been removed by natural selection because they had been deleterious. On the assumption of 60 000 genes, and 240 000 generations since human-chimpanzee divergence, they estimated a deleterious rate of 1.6 mutations per person per generation out of a total of 4.2 mutations per person generation.

The estimated deleterious mutation rates in chimpanzees and gorillas were very similar but those for rodent-specific lineages are about one order of magnitude less, possibly because of much smaller numbers of germ cell divisions in rodents. The very high deleterious rate in hominids may even be an underestimate. If the total gene number were 80 000 and an average coding sequence was 1800 nucleotides, the estimated deleterious mutation rate would be 2.5 mutations per person per generation and, on other grounds too, a more likely rate has been considered to be three deleterious mutations per person per generation (Crow, 1999).

So with three genetic deaths per person why are we not extinct? The data would suggest that harmful mutations need to be weeded out in clusters at a time. One way to achieve this would be if natural selection operated such that individuals with the most mutations are preferentially eliminated (e.g. harmful mutations interact). This could only happen in a sexual species where mutations are shuffled each generation by genetic recombination, and so the existence of such a high deleterious mutation rate has been taken as further vindication that sex (meiotic recombination) is an efficient way to eliminate harmful mutations.

Pathogenic Mutations are Preferentially Located at Certain Types of Intragenic DNA Sequence

Pathogenic mutations can occur at three types of DNA sequence at a gene locus.

- *The coding sequence of the gene.* This is where the great majority of recorded pathogenic mutations have been identified. Those due to nucleotide substitution are, in the vast majority of cases, nonsynonymous substitutions and mostly occur at first and second base positions of codons. However, very rarely, a synonymous codon substitution is not neutral as expected, but may cause disease by activating a cryptic splice site. Because of its relatively high mutability, the CpG dinucleotide is often located at hotspots for pathogenic mutation in coding DNA (Cooper and Youssoufian, 1988). Other hotspots include tandem repeats within coding DNA.

- *Intragenic noncoding sequences.* This is restricted to sequences which are necessary for correct expression of the gene, such as important intronic elements, notably the highly conserved GT and AG dinucleotides at the ends of introns, but also conserved elements of the untranslated sequences. Often such mutations represent a small component (~10–15%) of the total pathogenic mutations at a gene locus (Cooper *et al.*, 1995). However, in some disorders pathogenic splicing mutations may be common. In the case of the collagen disorder osteogenesis imperfecta they constitute a very common pathological mutation which is second in frequency only to substitutions leading to replacement of the highly conserved, structurally important glycine residues. The collagen genes have small exons and a comparatively large number of introns (often more than 50 and as many as 106 in the case of the *COL7A1* gene), making them exceptional targets for splicing mutations. Occasional pathogenic mutations have been recorded in the 52 UTR (such as in the case of hemophilia B Leyden) and appear to exert their effect at the transcriptional level. Several examples are also known of pathogenic mutations in the 32 UTR.
- *Regulatory sequences outside exons.* Most mutations located in regulatory sequences have been identified in conserved elements located just upstream of the first exon, notably promoter elements. In addition, other more distantly located regulatory elements may be sites of pathological mutation. For example, deletions which eliminate the â-globin LCR but leave the â-globin gene and its promoter intact result in almost complete abolition of â-globin gene expression and contribute to â-thalassemia. Clearly, in some cases a gene may be regulated by the product of a distantly related gene. For example, in the case of rare variants of α-thalassemia with mental retardation, the α-globin gene and its promoter may show no evidence of pathological mutation and the disease maps to an X-linked gene which encodes a transcription factor, one of whose target sequences is presumably the α-globin gene (Gibbons *et al.*, 1995).

The Mitochondrial Genome is a Hotspot for Pathogenic Mutations

Because of the very large size of the human nuclear genome, most mutations occur in nuclear DNA sequences. By comparison, the mitochondrial genome is a small target for mutation (about 1/200 000 of the size of the nuclear genome). Unlike nuclear genes, mitochondrial genes are present in numerous copies (there are thousands of copies of the mtDNA molecule in each human somatic cell; some cells, such as brain and muscle cells, have particularly high oxidative phosphorylation requirements and so more mitochondria). The mtDNA is inherited from the maternal oocyte, which is an exceptional cell with many more mtDNA molecules than somatic cells. Given that a mutation in mitochondrial DNA must arise on a single mtDNA molecule, one might intuitively expect that the chances of a single mtDNA mutation becoming *fixed* would be very low and the mutation rate correspondingly low. On these grounds, one could anticipate that the proportion of clinical disease due to pathogenic mutation in the mitochondrial genome should be extremely low. Instead, the frequency of 'mitochondrial disorders' is rather high and the mitochondrial genome can be considered to be a mutation hotspot. Different factors can explain this apparent paradox:

- *Differential target size for pathogenic mutation.* Pathogenesis is associated with mutations in coding DNA and the mitochondrial genome has a much higher percentage of coding DNA (93%) than found in the nuclear genome (3%). When this is taken into consideration, however, there is still a large imbalance: about 100 Mb of coding DNA in the nuclear genome but only 15.4 kb of coding sequence in the mitochondrial genome, giving a target ratio of 6000:1 in favour of the nuclear genome.

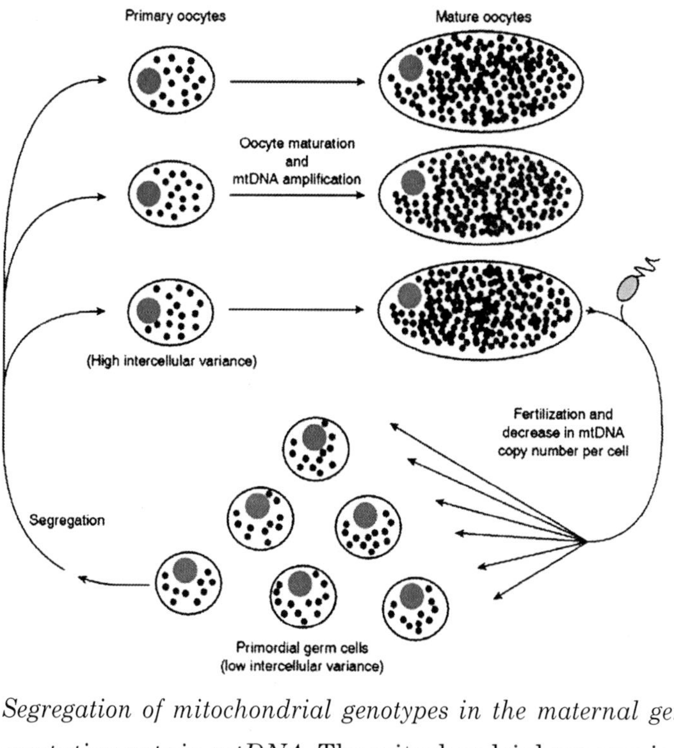

Figure: *Segregation of mitochondrial genotypes in the maternal germ line.*

- *High mutation rate in mtDNA.* The mitochondrial genome is much more prone to nucleotide change than the nuclear genome. Even although about 100 000 copies of the mitochondrial genome are maternally inherited in the fertilized oocyte there are mechanisms which permit rapid fixation of mutations in mitochondrial DNA. The combination of mtDNA instability and a high fixation rate means that the mutation rate in mitochondrial DNA is very high. Mutations have been reported to be fixed in the mitochondrial genomes of animal cells at a rate which is about 10 times greater than occurring in equivalent sequences in the nuclear genome (Brown *et al.*, 1979). This means that the small recombination-deficient animal mtDNA molecules appear to be evolving remarkably rapidly, corresponding to about 2–4% sequence divergence per million years. In contrast, plant mtDNA molecules are comparatively large (150 kb-2.5 Mb), have introns, engage in recombination and are evolving comparatively slowly.

The high instability of mtDNA has been postulated to result from several factors. The high rate of production of reactive oxygen intermediates by the respiratory chain is thought to cause substantial oxidative damage to mtDNA which, unlike nuclear DNA, is not protected by histones. The mtDNA also has to undergo many more rounds of replication than chromosomal DNA. Although several well-characterised mtDNA repair systems are now known, some frequent mutations cannot be repaired, including thymidine dimers.

Expression of Pathogenic Mutations

The degree to which a pathogenic mutation results in an aberrant phenotype depends on several factors:

- *The mutation class and the way in which the expression of the mutant gene is altered.* This may depend on the location of the mutation within the gene. Many pathogenic mutations result in abolition or substantial reduction of gene expression, but some lead to inappropriate expression. For example, overexpression of a gene product may cause an abnormal phenotype where gene dosage needs to be carefully regulated, and ectopic expression, that is expression in tissues where the gene is not normally expressed, may also be harmful.
- *The degree to which aspects of the aberrant phenotype are expressed in the heterozygote.* The presence of a single normal allele may be sufficient to maintain a clinically normal phenotype (as in recessively inherited disorders), or a milder phenotype when compared with that of mutant homozygotes, as in dominantly inherited disorders where the mutation is a simple *loss of function* mutation.
- *The degree to which expression of a mutant phenotype is influenced by other gene products.* The same mutant allele can have different phenotypic effects on different genetic backgrounds, depending on particular alleles at other gene loci (*modifier genes*).

- *The proportion and nature of cells in which the mutant gene is present.* Generally, mutations which are present in all the cells of an individual (inherited mutations) or in many of them (somatic mutations acquired very early in development) are likely to have a more profound effect than those present in a few cells (somatic mutations which arise at much later stages) or in cell types where the relevant gene is not expressed. Cancers, however, arise from unregulated division of cells produced from a single original mutant cell.
- *The parental origin of the mutation.* This is only known to be important in the case of the few genes which are *imprinted.*

Conserved Sequence Required for Normal Splicing, but some Occur in Sequences not Normally Required for Splicing

Many genes naturally undergo alternative forms of RNA splicing. In addition, mutations can sometimes produce an aberrant form of RNA splicing which is pathogenic. Sometimes this results in the sequences of whole exons being excluded from the mature RNA (exon skipping) or retention of whole introns. On other occasions, the abnormal splicing pattern may exclude part of a normal exon or result in new exonic sequences. Point mutations which alter a conserved sequence that is normally required for RNA splicing are comparatively common. Occasionally, however, aberrant splicing of a gene can be induced by mutation of other sequence elements which resemble splice donor or splice acceptor sequences but which are not normally involved in splicing (cryptic splice sites).

Mutations Which Alter Important Splice Site Signals

Often such mutations occur at the essentially invariant GT and AG dinucleotides located respectively at the start of an intron (*splice donor*) or at its end (*splice acceptor*). Flanking these important signals, however, are other conserved sequence elements which, if mutated, can also cause aberrant splicing. Mutations which alter such sequences can have different consequences:

- *Failure of splicing causing intron retention.* This can occasionally result, for example, when an intron is small and the neighbouring sequence lacks alternative legitimate splice sites or *cryptic splice sites* (sequences which resemble the consensus splice site sequences but which are not normally used by the splicing apparatus). The introduction of intronic sequence into the coding sequence of a mature mRNA will, at the very least, introduce additional amino acids and may cause a frameshift.
- *Exon skipping.* The splicing apparatus uses an *alternative* legitimate splice site. Mutation of a splice donor sequence results in skipping of

the upstream exon while mutation of the splice acceptor sequence results in skipping of the downstream exon. Often, the exclusion of an exon has a profound effect on gene expression: it may result in a frameshift, an unstable RNA transcript, or a nonfunctional polypeptide because of a loss of a critical group of amino acids.

In addition to the above, the *branch site* used in splicing may be mutated leading to defective splicing.

Mutations of Sequences Which are not Normally Important for RNA Splicing

Cryptic splice sites coincidentally resemble the sequences of authentic splice sites but are not normally used in splicing, unless a mutation alters the sequence so that the splicing apparatus now recognizes it as a normal splice site. The use of an intronic cryptic splice site will introduce new amino acids, while using an exonic cryptic splice site will result in a deletion of coding DNA. The former is a cautionary reminder that apparently silent mutations may yet be pathogenic. Note that in some cases mutations which occur within exons but not at cryptic splice sites can also induce skipping of that exon.

Mutations That Introduce a Premature Termination Codon Usually Result in Unstable mRNA but other Outcomes are Possible

Several different classes of mutation can introduce a premature termination codon (chain terminating mutations). Nonsense mutations produce a premature termination codon simply by substituting a normal codon with a stop codon. Frameshifting insertions and deletions usually also introduce a premature termination codon not too far downstream of the mutation site.

This happens because there is no selection pressure to avoid stop codons in the other translational reading frames and so given established nucleotide frequencies, at least one stop codon is usually encountered within a stretch of 100 nucleotides downstream of the mutation site. A variety of splice site mutations too can introduce a premature termination codon e.g. by skipping of a single exon containing a number of nucleotides that cannot be divided by 3. There are several possible consequences for gene expression for chain-terminating mutations:

- *Unstable mRNA.* This is by far the most frequent consequence. A mRNA carrying a premature codon is usually rapidly degraded *in vivo* by a form of RNA surveillance known as nonsense-mediated mRNA decay (Hentze and Kulozik, 1999; Culbertson, 1999). This can avoid the potentially lethal consequences of producing a truncated polypeptide which could interfere with vital cell functions.

- *Truncated polypeptide.* The production of a polypeptide truncated at the C terminus is a very rare outcome *in vivo* (the well-known protein truncation test which assays for mutations introducing a premature termination codon is carried out using an *in vitro* transcription-translation system). Nevertheless, some truncated polypeptides are produced *in vivo*. The effect on gene expression may be difficult to predict and will depend among other things on the extent of the truncation, the stability of the polypeptide product and its ability to interfere with expression of normal alleles.
- *Exon skipping.* Some nonsense mutations appear to induce skipping of constitutive exons *in vivo*. For example, a nonsense mutation in the middle of exon 51 of the *FBN1* fibrillin gene (corresponding to the C terminus of the protein) causes that exon to be skipped (Dietz *et al.*, 1993). As a result of exon skipping the abnormally spliced mRNA uses the normal stop codon and escapes nonsense-mediated mRNA decay unlike any full length mRNA which may be produced from the pre-mRNA. The abnormally spliced *FBN1* mRNA accumulates and is translated to give a dominant negative protein lacking C-terminal sequences.

The Pathogenic Potential of Repeated Sequences

The human genome, like other mammalian genomes, has a very high proportion of DNA sequences that are repeated. Tandem repeats in coding DNA include very short nucleotide repeats, moderately sized repeats and very large repeats that can include whole genes. Depending on the degree of sequence homology between the repeats, tandem repeats are liable to a variety of different genetic mechanisms causing sequence exchange between the repeats.

Often such sequence exchanges result in changes in the number of tandem repeats. A reduction in repeat number can often result in a pathogenic deletion, but expansion by sequence duplication can be pathogenic too. Certain chromosomal regions, notably the subtelomeric and pericentromeric regions, harbour large tracts of duplicated DNA and instability of such regions can predispose to disease (Eichler, 1998). Interspersed repeats can also cause pathogenic mutations by a different variety of mechanisms.

Slipped Strand Mispairing of Short Tandem Repeats Predisposes to Pathogenic Deletions and Frameshifting Insertions

Insertions and deletions in coding DNA are rare because they usually introduce a translational frameshift. However, occasionally, a series of tandem repeats of a small number of nucleotides occurs by chance in the coding sequence for a polypeptide. Such repeats, like classical microsatellite loci, are comparatively prone to mutation by slipped strand mispairing. As a result, the copy number of tandem repeats is liable to fluctuate, introducing a deletion or an insertion of one or more repeat units. If the mutation occurs in polypeptide-encoding DNA, a resulting deletion will often have a profound effect on gene expression.

Frameshifting deletions will normally result in abolition of gene expression. Even if the deletion does not produce a frameshift, deletions of one or more amino acids can still be pathogenic. Small frameshifting insertions will also be expected to lead to loss of gene expression and often the insertion is a tandem repeat of sequences flanking it. However, nonframeshifting insertions would often not be expected to be pathogenic, unless the insertion occurs in a critically important region, destabilising an essential structure or impeding gene function in some way. Note that large triplet repeat expansions can lead to disease by mechanisms that are not understood at present.

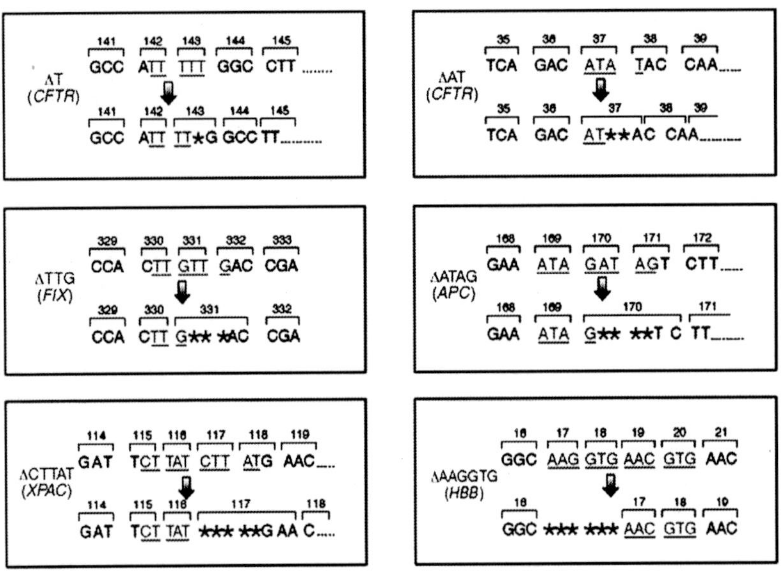

***Figure:** Short tandem repeats are deletion/insertion hotspots.*

Rapid Large-scale Expansion of Intragenic Triplet Repeats can Cause a Variety of Diseases but the Mutational Mechanism is Not well Understood

Sometimes microsatellites within or in the immediate vicinity of a gene can expand to considerable lengths and affect gene expression, causing disease. In some cases, a modestly expanded repeat which causes disease may be perfectly stable and be propagated without change in size through several generations. For example, triplet repeat expansion leading to long polyalanine tracts in the *HOXD13* gene cause a form of synpolydactyly, probably as a result of unequal crossover (Warren, 1997), but the expanded repeat is stable (Akarsu *et al.*, 1996). In other cases, however, the expanded triplet repeat is *unstable* and the discovery that human disease can be caused by large-scale expansion of highly unstable trinucleotide repeats was quite unexpected.

Studies in other organisms had not revealed precedents for such a phenomenon, but the list of human examples is now considerable. In addition to unstable triplet repeat expansion, the majority of disease alleles at the cystatin B gene which cause progressive myoclonus epilepsy involve expansions of a 12 nucleotide repeat (CCCCGCCCCGCG; Lalioti *et al.*, 1998).

Tandem trinucleotide repeats are not infrequent in the human genome. Although there are 64 possible trinucleotide sequences, when allowance is made for cyclic permutations $(CAG)_n = (AGC)_n = (GCA)_n$ and reading from either strand 2 $(CAG)_n$ on one strand = 52 $(CTG)_n$ on the other], there are only 10 different trinucleotide repeats.

Most of these are known as usefully polymorphic microsatellite markers but, in addition, certain repeats show anomalous behaviour which can cause abnormal gene expression.

In each case, repeats below a certain length are stable in mitosis and meiosis while, above a certain threshold length, the repeats become extremely unstable. These unstable repeats are virtually never transmitted unchanged from parent to child. Both expansions and contractions can occur, but there is a bias towards expansion. The average size change often depends on the sex of the transmitting parent, as well as the length of the repeat.

The ten possible trinucleotide repeats. Both DNA strands are shown. All other trinculeotide repeats are cyclic permutations of one or another of these.

- *Genes which show modest expansions of* $(CAG)_n$ *repeats within the coding sequence.* Typically, the stable and nonpathological alleles have 10–30 repeats, while unstable pathological alleles have modest expansions, often in the range of 40–200 repeats. Transcription and translation of the gene are not affected by the expansion. The resulting protein product shows a gain of function: its long polyglutamine tract causes it to aggregate within certain cells and kill them.
- *Genes which show very large expansions of a noncoding repeat.* For some genes, various types of triplet repeat (e.g. CGG, CCG, CTG, GAA) found in the promoter, the untranslated regions or intronic sequences can undergo very large expansions in such a way as to inhibit gene expression, causing loss of function. Typically, the stable and nonpathological alleles have 5–50 repeats, while unstable pathological alleles have several hundreds or thousands of copies.

Intergenerational changes are normally reported as parent-child comparisons of blood lymphocyte DNA. There is little information about when in gametogenesis, fertilization or embryogenesis the changes arise. Limited studies of sperm show that highly expanded *DM* and *FRAXA* (fragile-X syndrome) repeats are not transmitted by affected males, although modest expansions can be. The largest expansions in Huntington disease (which, however, are small compared with large *FRAXA* or *DM* expansions) are seen in sperm, consistent with the observation that the severest cases inherit the disease from their father. At least for *FRAXA, DM* and Kennedy disease, the expanded repeats are mitotically unstable, so that a blood sample shows a smear of heterogeneous expanded repeats sizes.

However, *in vitro*, even large repeats are stable. Thus, whatever the mechanism, it is not operative in all cells. The basis of the unstable expansions is very largely unknown and this type of mutagenic event has not been identified thus far in genetically tractable organisms such as *E. coli*, yeast or *Drosophila.* There is also evidence that the unstable expansion mechanism may not have a parallel in some other mammals such as mice. Human transgenes containing long trinucleotide repeats show virtually no instability after being propagated through several generations in transgenic mice whereas the same sequences may show a 100% probability of expansion when transmitted in the human germline (Djian, 1998).

Investigations have also suggested that arrays of triplet repeats may be able to form alternative DNA structures, such as DNA hairpins, triplex DNA and quadruplex DNA (Sinden, 1999) but their significance if formed *in vivo* is unknown.

The repeats have also been envisaged as possible protein-binding sites and protein-binding at the RNA level has also been envisaged to contribute to pathogenesis in some cases, notably in myotonic dystrophy (Philips *et al.,* 1998). Slipped strand mispairing is likely to be a component of the expansion mechanism, given the observation that interrupted repeats appear to be stable and only homogeneous repeats are unstable.

For example, in spinocerebellar ataxia type 1, 123/126 normal sized CAG repeats were interrupted by one or two CAT triplets, while 30/30 expanded alleles contained no interruption. One problem with all these mispairing mechanisms is that they should result in contractions as well as expansions and this is not seen. Instead, after a certain threshold size, there appears to be a clear bias towards continued expansion of the size of the repeat unit array. Because understanding of trinucleotide repeats is progressing very rapidly at the time of writing, the reader is advised to consult a recent review for more information.

Tandemly Repeated and Clustered gene Families may be Prone to Pathogenic Unequal Crossover and Gene Conversion-like Events

Many human and mammalian gene clusters contain nonfunctional pseudogenes which may be closely related to functional gene members. Interlocus sequence exchanges between pseudogenes and functional genes can result in disease by removing or altering some or all of the sequence of a functional gene. For example, unequal crossover (or unequal sister chromatid exchange) between a functional gene and a related pseudogene can result in deletion of the functional gene or the formation of fusion genes containing a segment derived from the pseudogene. Alternatively, the pseudogene can act as a donor sequence in gene conversion events and introduce deleterious mutations into the functional gene.

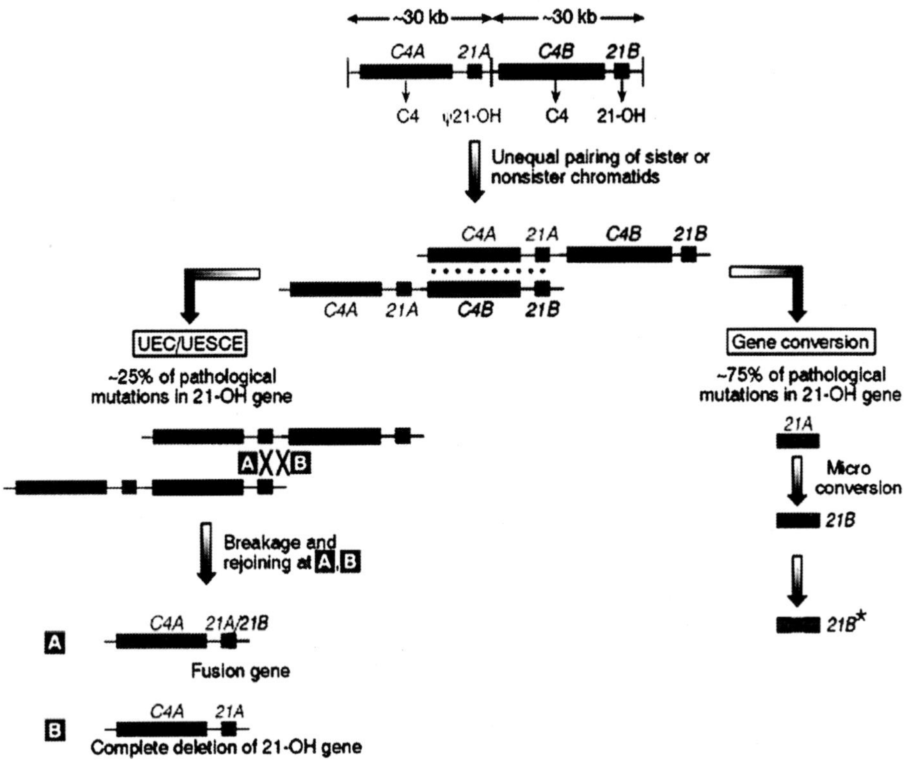

Figure: *Almost all 21-hydroxylase gene mutations are due to sequence exchange with a closely related pseudogene.*

The classical example of pathogenesis due to gene- pseudogene exchanges is steroid 21-hydroxylase deficiency, where over 95% of pathogenic mutations arise as a result of sequence exchanges between the functional 21-hydroxylase gene, *CYP21B*, and a very closely related pseudogene, *CYP21A*. The two genes occur on tandemly repeated DNA segments approximately 30 kb long which also contain other duplicated genes, notably the complement C4 genes, *C4A* and *C4B*. Large pathogenic deletions uniformly result in removal of about 30 kb of DNA, corresponding to one repeat unit length, and analysis of *de novo* 21-hydroxylase deficiency mutations has provided strong evidence for pathogenic deletions arising as a result of meiotic unequal crossover (Sinnott *et al.,* 1990).

Virtually all of the 75% of pathogenic point mutations are copied from deleterious mutations in the pseudogene, suggesting a gene conversion mechanism. Analysis of one such mutation which arose *de novo* suggests that the conversion tract is a maximum of 390 bp (Collier *et al.,* 1993).

Gene conversion events are also found in the duplicated C4 genes, both of which are normally expressed. A likely priming event for conversions in the *CYP21-C4* gene cluster is unequal pairing of chromatids so that a *CYP21A-C4A* unit pairs with a *CYP21B-C4B* unit.

Interspersed Repeats often Predispose to Large Deletions and Duplications

In several cases, the endpoints of deletions are marked by very short direct repeats. For example, the breakpoints in numerous pathogenic deletions in mtDNA occur at perfect or almost perfect short direct repeats.

Of these, the most common is a deletion of 4977 bp which has been found in multiple patients with Kearns-Sayre syndrome, an encephalomyopathy characterised by external ophthalmoplegia, ptosis, ataxia and cataract.

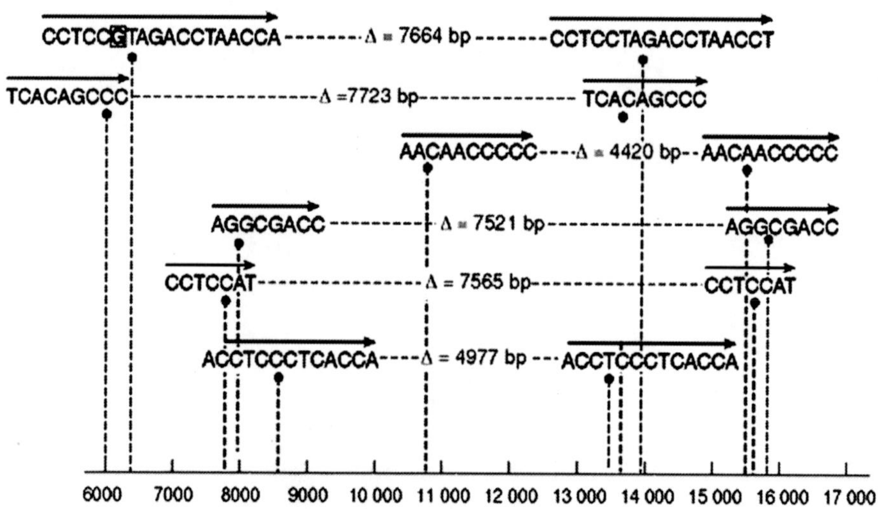

Figure: *Short direct repeats mark the endpoints of many pathogenic deletions in the mitochondrial genome.*

The deletion results in elimination of the intervening sequence between two perfect 13 bp repeats and loss of the sequence of one of the repeats. The mitochondrial genome is recombination-deficient and Shoffner *et al.* 1989 have postulated that such deletions arise by a replication slippage mechanism, similar to that occurring at short tandem repeats. Partial duplications of the mitochondrial genome are also distinctive features of certain diseases, notably Kearns-Sayre syndrome. The ends of the duplicated sequences, like those of the common deletions, are often marked by short direct repeats, and the mechanisms of duplication and deletion appear to be closely related (Poulton and Holt, 1994).

The Alu Repeat as a Recombination Hotspot

Some large-scale deletions and insertions may be generated by pairing of nonallelic interspersed repeats, followed by breakage and rejoining of chromatid fragments. For example, the *Alu* repeat occurs approximately once every 4 kb and mispairing between such repeats has been suggested to be a frequent cause of deletions and duplications. Some large genes have many internal *Alu* sequences in their introns or untranslated sequences, making them liable to frequent internal deletions and duplications.

For example, the 45-kb low density lipoprotein receptor gene has a relatively high density of *Alu* repeats (approximately one every 1.6 kb). A very high frequency of pathogenic deletions in this gene are likely to involve an *Alu* repeat, usually at both endpoints, and occasional pathogenic intragenic duplications also involve *Alu* repeats (Hobbs *et al.*, 1990). Such observations have suggested a general role for *Alu* sequences in promoting recombination and recombination-like events. Initial gene duplications in the evolution of clustered multigene families may often have involved an unequal crossover event between *Alu* repeats or other dispersed repetitive elements. It should be noted, however, that some *Alu*-rich genes do not appear to be loci for frequent *Alu*-mediated recombination.

Pathogenic Inversions can be Produced by Intrachromatid Recombination Between Inverted Repeats

Occasionally, clustered inverted repeats with a high degree of sequence identity may be located within or close to a gene. The high degree of sequence similarity between inverted repeats may predispose to pairing of the repeats by a mechanism that involves a chromatid bending back upon itself. Subsequent chromatid breakage at the mispaired repeats and rejoining can then result in an inversion, in much the same way as the natural mechanism used for the production of some immunoglobulin ê light chains.

The classic example of pathogenic inversions is a mutation which accounts for more than 40% of cases of severe hemophilia A. Intron 22 of the factor VIII gene, *F8*, contains a CpG island from which two internal genes are transcribed: *F8A* in the opposite direction to the host gene *F8*, and *F8B* in the same direction as *F8*. *F8A* belongs to a gene family with two other closely related members located several hundred kilobases upstream of *F8* gene and transcribed in the opposite direction to *F8A*. As a result, the region between the *F8A* gene and the other two members is susceptible to inversions - the *F8A* gene can pair with either of the other two members on the same chromatid, and subsequent chromatid breakage and rejoining in the region of the paired repeats results in an inversion which disrupts the factor VIII gene.

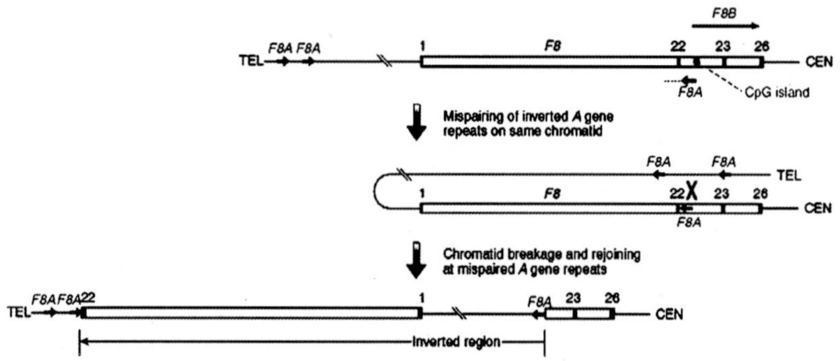

***Figure:** Inversions disrupting the factor VIII gene result from intrachromatid recombination between inverted repeats.*

DNA Sequence Transposition is not Uncommon and can Cause Disease

A proportion of moderately and highly repeated interspersed elements are capable of transposition via an RNA intermediate. Defective gene expression due to DNA transposition is comparatively rare and represents only a small component of molecular pathology. However, several examples have been recorded of genetic deficiency due to insertional inactivation by retrotransposons. For example, in one study, hemophilia A was found to arise in two out of 140 unrelated patients as a result of a *de novo* insertion of a LINE-1 (*Kpn*) repeat into an exon of the factor VIII gene (Kazazian *et al.,* 1988). Other instances are known of insertional inactivation by an actively transposing *Alu* element, as in a case of neurofibromatosis type 1 (Wallace *et al.*, 1991). Additionally, a number of other examples have been recorded of pathogenesis due to intragenic insertion of undefined DNA sequences.

DNA Repair

- DNA in cells suffers a wide range of damage:
- purine bases are lost by spontaneous fission of the base-sugar link;
- cytosines, and occasionally adenines, spontaneously deaminate to produce uracil and hypoxanthine respectively;

- many chemicals, for example alkylating agents, form adducts with DNA bases;
- reactive oxygen species in the cell attack purine and pyrimidine rings
- ultraviolet light causes adjacent thymines to form a stable chemical dimer;
- mistakes in DNA replication result in incorporation of a mismatched base;
- ionising radiation causes single- or double-strand breaks;
- mistakes in replication or recombination leave strand breaks in DNA.

All these lesions must be repaired if the cell is to survive. The importance of effective DNA repair systems is highlighted by the severe diseases affecting people with deficient repair systems.

DNA Repair Usually Involves Cutting out and Resynthesising a Whole area of DNA Surrounding the Damage

To cope with all these forms of damage, cells must be capable of several different types of DNA repair. DNA repair seldom involves simply undoing the change that caused the damage. Almost always a stretch of DNA containing the damaged nucleotide(s) is excised and the gap filled by resynthesis. There are at least five main types of DNA repair in human cells:

- *Direct repair* reverses the DNA damage. A specific enzyme is able to dealkylate O^6-alkyl guanine directly. In bacteria thymine dimers can be removed in a photoreactivation reaction that depends on visible light and an enzyme, photolyase. Mammals possess enzymes related to photolyase, but use them for a quite different purpose, to control their circadian clock (Van der Horst *et al.,* 1999).
- *Base excision repair (BER)* uses glycosidase enzymes to remove abnormal bases. An endonuclease, AP endonuclease, cuts the sugar-phosphate backbone at the position of the missing base. A few nucleotides of the DNA strand are stripped back by exonucleases, the gap is filled by resynthesis, and the remaining nick is sealed by DNA ligase III. The same process is used to repair spontaneous depurination. Interestingly, no human diseases caused by defective BER are known. Maybe any such defect would be lethal, since BER corrects much the commonest type of DNA damage.
- *Nucleotide excision repair (NER)* removes thymine dimers and large chemical adducts. Defects in NER cause the autosomal recessive disease xeroderma pigmentosum (Lambert *et al.,* 1998). Seven complementation groups, XPA-XPG, have been defined by cell fusion studies. XP patients are exceedingly sensitive to UV light. Sun-exposed skin develops thousands of freckles, many of which progress to skin cancer.

- *Post-replication repair* is required to correct double-strand breaks (Haber, 1999). The usual mechanism is a gene conversion-like process (recombinational repair), where a single strand from the homologous chromosome invades the damaged DNA helix. Alternatively, broken ends are rejoined regardless of their sequence, a desperate measure that is likely to cause mutations. The eukaryotic machinery for recombination repair is less well defined than the excision repair systems. Human genes involved in this pathway include *NBS* (mutated in Nijmegen breakage syndrome, MIM 251260), *BLM* (mutated in Bloom syndrome, MIM 210900) and the *BRCA2* and maybe *BRCA1* breast cancer susceptibility genes.
- *Mismatch repair* corrects mismatched base pairs caused by mistakes in DNA replication. Cells deficient in mismatch repair have mutation rates 100–1000 times higher than normal, with a particular tendency to replication slippage in homopolymeric runs. In humans the mechanism involves at least five proteins and defects cause hereditary nonpolyposis colon cancer.

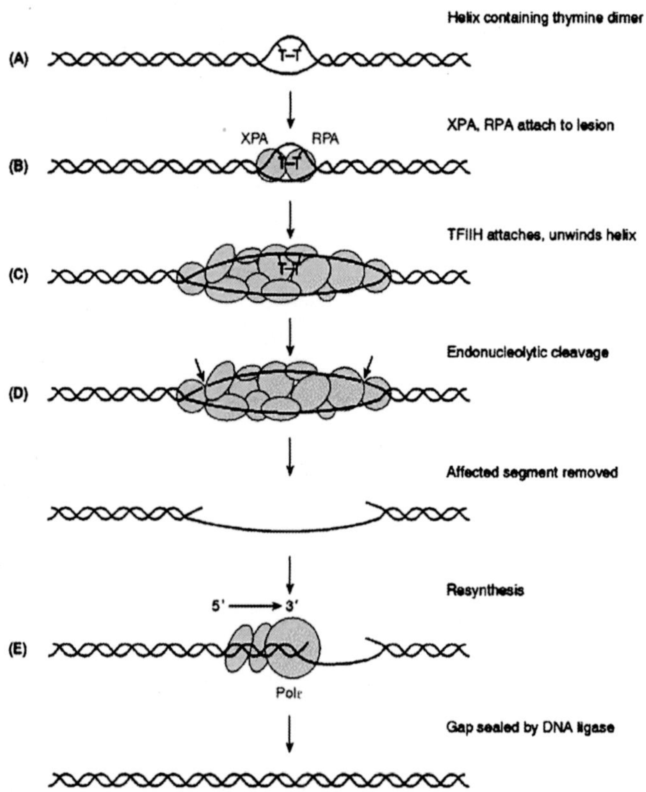

Figure: *A possible scheme for nucleotide excision repair in humans.*

All these systems, except for direct repair, require exo and endonucleases, helicases, polymerases and ligases, usually acting in multiprotein complexes that have some components in common. Sorting out the individual pathways has been greatly aided by the very strong conservation of repair mechanisms across the whole spectrum of life. Not only the reaction mechanisms but also the protein structures and even gene sequences are often conserved from *E. coli* to man. A downside of the conservation is a confusing gene nomenclature, referring sometimes to human diseases (*XPA* etc.), sometimes to yeast mutants (*RAD* genes) and sometimes to mammalian cell complementation systems (ERCC = excision repair cross-complementing) - for example *XPD*, *ERCC2* and *RAD3* are the same gene in man, mouse and yeast. Generally eukaryotes have multiple proteins corresponding to each single protein in *E. coli*, so that, for example, nucleotide excision repair requires six proteins in *E. coli* but at least 30 in mammals.

DNA Repair Systems Share Components and Processes

As well as sharing components with each other, many repair systems share components with the machinery for DNA replication, transcription and recombination. DNA polymerases and ligase are required for both DNA replication and resynthesis after excision of a defect. The recombination machinery is involved in double-strand break repair. The link with transcription is particularly intriguing (Lehmann, 1995). The general transcription factor TFIIH is a multiprotein complex that includes the XPB and XPD proteins. TFIIH exists in two forms. One form is concerned with general transcription and the other with repair, probably specifically repair of transcriptionally active DNA. This system is deficient in two rare diseases, Cockayne syndrome (MIM 216400) and trichothiodystrophy (MIM 601675). Clinically and in cell biology, CS and TTD both overlap XP, and in some cases the same genes are responsible, but CS and TTD patients have developmental defects that presumably reflect defective transcription, and they do not have the cancer susceptibility of XP patients.

Hypersensitivity to Agents that Damage DNA

Many human diseases that involve hypersensitivity to DNA-damaging agents, or a high level of cellular DNA damage, are not caused by defects in the DNA repair systems themselves, but by a defective cellular response to DNA damage. Normal cells react to DNA damage by stalling progress through the cell cycle at a checkpoint until the damage has been repaired, or triggering apoptosis if the damage is irrepairable. Part of the machinery for doing this involves the ATM protein.

The role of ATM. Briefly, it senses DNA damage and relays the signal to the p53 protein, the 'guardian of the genome'. People with no functional

ATM have ataxia telangiectasia (MIM 208900; Lambert *et al.*, 1998). Their cells are hypersensitive to radiation, and they have chromosomal instability and a high risk of malignancy, but the DNA repair machinery itself is intact. Fanconi anemia (MIM 227650) is another heterogeneous group of diseases (at least five complementation groups) marked by defective responses to DNA damage, without specific defects in DNA repair.

Physical and Transcript Mapping

A wide variety of physical mapping strategies have been used to analyse the DNA of complex eukaryotic genomes. In the following sections, an arbitrary distinction is made between two classes of physical mapping:

- low resolution physical mapping - the smallest map unit that can be resolved is typically one to several megabases of DNA;
- high resolution physical mapping - the resolution is typically very high, from hundreds of kilobases to a single nucleotide.

Since mammalian DNA has only a very small percentage of coding DNA (~3% in the case of the human genome), a variety of transcript mapping methods have been developed for selectively identifying and studying transcribed sequences, the most important and, arguably, most interesting component of the genome. 1Low resolution physical mapping

Somatic Cell Hybrid Panels can Permit Chromosomal Localisation of any Human DNA Sequence

Under certain experimental conditions, cultured cells from different species can be induced to fuse together, thereby generating somatic cell hybrids. In human genetic mapping, hybrid cells are typically constructed by fusing human cells and mouse or hamster cells. The initial fusion products are described as heterokaryons because the cells contain both a human and a rodent nucleus. Eventually, heterokaryons proceed to mitosis, and the two nuclear envelopes dissolve. Thereafter, the human and rodent chromosomes are brought together in a single nucleus. The hybrid cells are unstable. For reasons that remain unknown, most human chromosomes fail to replicate in subsequent rounds of cell division and are lost. This gives rise, eventually, to a variety of more-or-less stable hybrid cell lines, each with the full set of rodent chromosomes plus a few human chromosomes. The loss of the human chromosomes occurs essentially at random, but can be controlled by selection.

The human chromosomes in somatic cell hybrids can conveniently be identified by PCR screening with sets of chromosome-specific primers (Abbott and Povey, 1991). By collecting hybrid cell lines with different human chromosome contents it is possible to generate a hybrid cell panel that can be used to map any human DNA sequence to a specific chromosome. To do this, each of the hybrid cell lines is tested for the presence of the human sequence of interest.

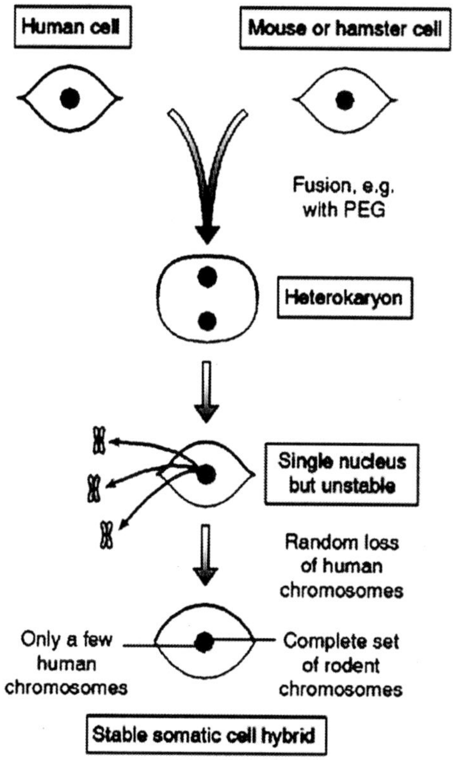

Figure: *Fusion of cells from different species can result in stable somatic cell hybrids.*

A PCR assay can be used with primers specific for that sequence (this is a form of STS content mapping) or the relevant DNA sequence can be labelled and used as a hybridisation probe. Both of these methods require that the human sequence should be distinguishable from any homologous rodent sequence. Somatic cell hybrids could also be used to establish which chromosome encodes a human enzyme or other gene product by biochemical typing of the hybrids. Whichever human chromosome contains the sequence of interest, it will be present in all the cell lines that are positive for the test and absent from all cell lines that are negative for the test.

Monochromosomal hybrids generated by microcell fusion allow direct and rapid chromosomal localisation of human DNA clones. Traditional somatic cell hybrids generally contain several human chromosomes. Microcell fusion (Fournier and Ruddle, 1977) can be used to generate monochromosomal hybrids that contain only a single human chromosome.

The donor human cells are subjected to prolonged mitotic arrest by continued exposure to an inhibitor of mitotic spindle formation, such as colcemid. This causes the chromosome content of a cell to become partitioned into discrete subnuclear packets (micronuclei). Centrifugation in the presence of cytochalasin B (a mitotic spindle inhibitor), results in the formation of microcells.

Microcells consist of a single micronucleus with a thin rim of cytoplasm, surrounded by an intact plasma membrane. In much the same way as for normal donor cells, the microcells can be fused with recipient rodent cells to generate hybrids (microcell mediated chromosome transfer). Some microcell hybrids contain a few donor chromosomes, but the simplest contain a single donor chromosome.

Subchromosomal Mapping

Monochromosomal human-rodent hybrid cell lines have been established for each of the human chromosomes (Cuthbert *et al.,* 1995). If the sequence of a particular human DNA is known, a PCR assay can be established and used to type a complete monochromosomal hybrid panel, thereby establishing the chromosomal location quickly. Alternatively, if no DNA sequence is available, a human fragment of interest can still be labelled and used as a hybridisation probe against a Southern blot of genomic DNA samples from a monochromosomal hybrid panel.

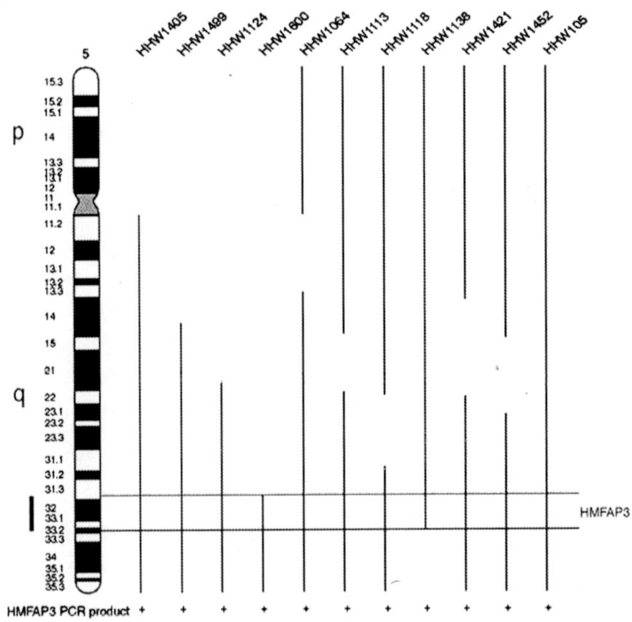

Figure: Subchromosomal localisation can be achieved by mapping against a panel of hybrid cells containing translocation or deletion chromosomes.

Conventional somatic cell hybrids are a relatively crude tool for physical mapping. More refined mapping is possible using hybrids that contain only part of a particular human chromosome. Translocation hybrids and deletion hybrids are made using donor human cells that have a chromosomal translocation or deletion.

To be useful, the hybrids must lack the normal homolog of the chromosome of interest. Such hybrids can be used for subchromosomal mapping of a human sequence-tagged site or biochemical marker. They are especially useful for defining the sequences removed by microdeletions, by segregating the deletion-carrying chromosome away from its normal homolog.

Human DNA Clones can be Physically Mapped to any Desired Resolution

A more general approach to producing tools for physical mapping involves artificially breaking human chromosomes and transferring the fragments into rodent cells.

Chromosome-mediated Gene Transfer

One of the first techniques to use this approach was chromosome-mediated gene transfer (CMGT). Fragments of purified mitotic chromosomes from a donor, such as a human fibroblast, are coprecipitated with calcium phosphate on to the surface of a recipient rodent cell line in monolayer culture.

Human chromosome fragments enter the recipient cells, such as mouse fibroblasts, and integrate into the chromosomes, resulting in stable transformation. As a result, hybrids can be established that retain segments of human DNA (transgenomes) of a size that is useful for mapping (usually in the range of 1–50 Mb). However, the transgenomes are prone to frequent re-arrangements, so CMGT is more suited to functional assays of complex loci than as a mapping tool.

Irradiation Fusion gGene Transfer

The most valuable hybrids for gene mapping are radiation hybrids. Donor cells are subjected to a lethal dose of radiation which fragments their chromosomes. The average size of a fragment is a function of the dose of radiation. After irradiation the donor cells are fused with recipient cells of a different species. A selection system is used to pick out recipient cells that have taken up some of the donor chromosome fragments.

These cells are useful for mapping insofar as they have taken up a random set of other chromosome fragments from the donor, as well as the selected fragment. Stably incorporated donor fragments are either integrated into rodent chromosomes or are assembled into novel human minichromosomes formed around fragments containing a functional centromere. Although this procedure was first proposed by Goss and Harris in 1975, it was not used seriously until 1990, when hybrids were constructed using irradiated monochromosomal hybrid cells as donors (Cox *et al.*, 1990). When a set of DNA markers from the human chromosome is assayed in a panel of such radiation hybrids, the patterns of cross-reactivity can be used to construct a map. The frequency of breakage between two markers can be defined by a value θ, analogous to the recombination frequency in meiotic mapping. θ varies from 0 (the two markers are never separated) to 1.0 (the two markers are always broken apart).

As in meiotic mapping, θ underestimates the distance between markers that are far apart on the same chromosome, in this case because a cell can take up two markers on separate fragments. A more accurate estimate is provided by a mapping function, $D = -\ln(1 - \theta)$, which is analogous to the Haldane mapping function used in meiotic linkage analysis. D is measured in centirays (cR). D is dependent on the dosage of radiation, so it is referenced against the number of rads. For example, a distance of 1 cR_{8000} between two markers represents a 1% frequency of breakage between them after exposure to 8000 rad of X-rays.

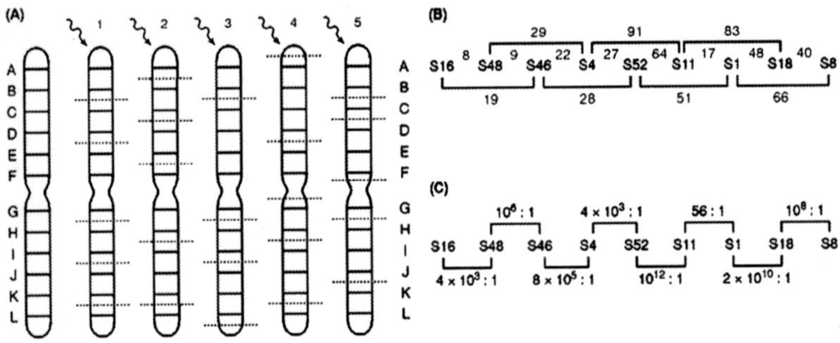

Figure: Constructing radiation hybrid maps.

Radiation hybrids derived from monochromosomal hybrid donor cells have been superseded by whole-genome radiation hybrids where the donor is an irradiated normal human diploid cell. The first such panel consisted of 199 hybrids made by fusing an irradiated 46,XY human fibroblast cell line to TK^- hamster cells (Walter *et al.*, 1994). Gyapay *et al.* (1996) used 404 microsatellite markers of known location to show that this hybrid panel could generate accurate maps, and then used it to map 374 unmapped ESTs.

A subset of 93 of the hybrids has been made widely available as the Genebridge 4 panel. The 93 hybrids average 32% retention of any particular human sequence, with an average fragment size of 25 Mb. Laboratories can map any unknown STS by scoring the 93 Genebridge hybrids and comparing the pattern with patterns of previously mapped markers held on a central server.

This has turned into an extremely powerful and convenient tool for physically mapping any STS or EST. A second human-hamster panel, Stanford G3, was made using a higher dose of radiation, so that the average human fragment size is smaller. The 83 hybrids in G3 average 16% retention of the human genome, with an average fragment size of 2.4 Mb. Thus G3 can be used for finer mapping.

Chromosomal in Situ Hybridisation has been Revolutionised by Fluorescence in Situ Hybridisation Techniques

Chromosomal *in situ* hybridisation typically uses an air-dried microscope slide preparation of metaphase chromosomes, in which the chromosomal DNA has been denatured by exposure to formamide before hybridisation with a suitable probe. Isotopes such as ^{32}P are not suitable for probe labelling because their high energy causes scattering of the signal.

Instead, isotopes with weaker emission of beta radiation have been used. Owing to the high background noise when using such isotopes, however, the genuine signal is not easily identified and complex statistical analysis is required to differentiate genuine signal from background noise.

More recently, the sensitivity and resolution of *in situ* hybridisation has been increased significantly by the development of fluorescence *in situ* hybridisation (FISH). In this technique, the DNA probe is either labelled directly by incorporation of a fluorescent labelled nucleotide precursor, or indirectly by incorporation of a nucleotide containing a reporter molecule (such as biotin or digoxigenin) which after incorporation into the DNA is then bound by fluorescently labelled affinity molecule.

To increase the intensity of the hybridisation signal, large DNA probes are preferred, usually cosmid clones containing around 40 kb of insert. Because such large sequences will contain many interspersed repetitive DNA sequences, it is necessary to use chromosome *in situ* suppression hybridisation (Lichter *et al.,* 1990). Essentially, this is a form of competition hybridisation: before the main hybridisation, the probe is mixed with a large excess of unlabelled total genomic DNA and denatured, thereby saturating the repetitive elements in the probe, so that they no longer mask the signal generated by the unique sequences.

FISH has the advantage of providing rapid results which can be scored conveniently by eye using a fluorescence microscope. In metaphase spreads, positive signals show as double spots, corresponding to probe hybridised to both sister chromatids. Using sophisticated image processing equipment and reporter-binding molecules carrying different fluorophores, it is possible to map and order several DNA clones simultaneously. The maximum resolution of conventional FISH on metaphase chromosomes is several megabases.

The use of the more extended prometaphase chromosomes can permit 1 Mb resolution but, because of problems with chromatin folding, two differentially labelled probe signals may appear to be side-by-side, unless they are separated by distances greater than 2 Mb. Recently, however, new variations have been developed, permitting very high resolution. There are numerous applications for chromosome FISH but a special one is chromosome painting which has permitted a form of molecular karyotyping.

Flow Cytometry Permits Sorting of Individual Chromosomes and the Construction of Chromosome-specific DNA Libraries

Human chromosomes show considerable variation in size and DNA content. In addition, the base composition can vary considerably: chromosomes that are known to be gene-rich or gene-poor have a comparatively high % (G + C) or % (A + T) respectively. For example, chromosomes 21 and 22 are similar in size and in DNA content, but chromosome 21 is gene-poor and has a relatively high % (A + T), whereas chromosome 22 is gene-rich and has a relatively high % (G + C). As a result of the size differences and differences in base composition, human chromosomes can be separated by flow sorting (also called flow cytometry). In this technique, chromosome preparations are stained with a DNA-binding dye which can fluoresce in a laser beam.

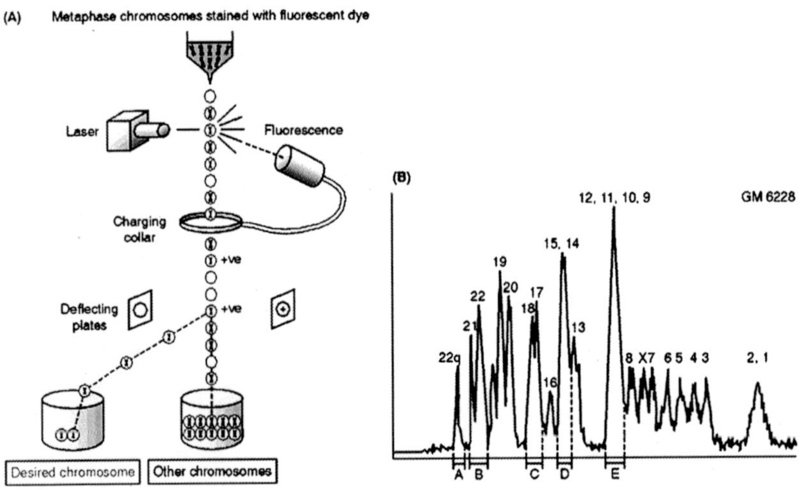

Figure: Fractionating chromosomes in a flow cytometre. (A) Principle of chromosome fractionation.

The amount of fluorescence exhibited by a given chromosome is proportional to the amount of dye bound, which in turn is largely proportional to the amount of DNA, and hence the size of the chromosome. Additionally, certain dyes, such as chromomycin A_3, show preferential binding to GC-rich sequences, while others, such as Hoechst 33258, bind preferentially to AT-rich regions.

Because individual chromosomes can bind different amounts of fluorescent dyes, they can be fractionated in a flow cytometre. A stream of droplets containing stained chromosomes is passed through a finely focused laser beam at a rate of about 1000–2000 chromosomes per second, and the fluorescence of individual chromosomes present in individual droplets is monitored by a photomultiplier. The intensity of fluorescence of the different chromosomes can then be recorded. By using sort windows to define a range of fluorescence intensities corresponding to a particular type of chromosome, droplets containing the desired chromosome can be given an electric charge and deflected from the main stream on to a separate collecting grid.

The resulting flow karyogram may show good separation of some human chromosomes; others such as chromosomes 9–12 are not well separated. However, monochromosomal somatic cell hybrids can be used instead, and often result in good separation of the human chromosome. As a result, panels of chromosome-specific DNA representing each of the 24 human chromosomes have been obtained.

These can be used to obtain a chromosomal localisation for any human DNA fragment: the fragment is labelled and hybridised to dot-blots of DNA from the 24 chromosomes ('flow-blots'), or a PCR assay can be used, if appropriate. Additionally, purified DNA from a specific type of chromosome can be amplified by cell-based or PCR-based DNA cloning to generate chromosome-specific DNA libraries (Davies et al., 1981). Chromosome libraries constructed using cosmid vectors have been particularly useful because YACs from a specific chromosome can be related quickly to cosmid clones from the same region. In addition, it has been possible in some cases to construct human chromosome-specific YAC libraries.

Clone Gridding and Increasing Automation Have Facilitated Large-scale Physical Mapping Efforts

Using current technologies, physical mapping of complex genomes is arduous and continues to be hampered by various practical difficulties:

- Chimeric DNA clones. Cloning artifacts can mean that cells contain two noncontiguous pieces of DNA from the desired genome. This may occur as a result of *co-ligation* of two different restriction fragments, internal deletion and *cotransformation*. Yeast cells that have been transformed by YACs and used to propagate have often been much more recombination proficient than typical *E. coli* hosts, and so YACs are often chimeric.
- Repetitive DNA. Arranging clones in a physical order means relying on some clone-specific features. DNA sequences that are repeated can make the task of ordering clones a difficult one, and may cause problems in sequencing DNA.

- Unclonable sequences. Some DNA sequences, including classes of repetitive DNA that are not well represented in bacterial and yeast cells, may not easily be propagated in such cells. A variety of different genomic libraries may need to be screened (including those constructed with vectors containing a low copy number origin of replication such as BAC and P1 libraries).

A major requirement for efficient physical mapping of complex genomes is the need for gridded DNA libraries. Following transformation of the host cells, individual bacterial or yeast colonies are picked into individual wells of microtiter dishes and stored in numerous multiwell dishes. As a result, each cell clone can be assigned an identifying grid coordinate, comprising dish number, row number and column number. When libraries need to be screened by hybridisation with a labelled DNA probe, it is again most efficient to have previously prepared high density gridded colony filters in which the colonies have been spotted on to the appropriate membrane in a gridded fashion with a uniform number of rows and columns.

This simplifies identification of hybridisation-positive colonies. More importantly, it facilitates integration of the results from numerous independent research groups. This is possible because of increasing automation. Robots have been designed to pick and spot colonies, and numerous copies of reference libraries can be distributed, either as cell cultures or, more conveniently, as sets of colony filters and sent to researchers throughout the world. Hybridisation data can then be sent back from individual laboratories to the distributing centres and the fixed geometry of the library coordinates simplifies the task of integrating the information received.

The Human Genome Project is one of several projects to benefit from such technologies, and sets of hybridisation filters containing gridded colonies for total genome YAC libraries, chromosome-specific cosmid libraries and other libraries have been distributed extensively. In addition, pools of yeast colonies have been distributed in agarose plugs to permit PCR typing for a sequence of interest. Again, this is done in a hierarchical coordinate manner to simplify the screening of libraries.

Clone Contigs can be Assembled Rapidly over Large Fractions of a Genome by Random clone Fingerprinting

Chromosome walking is a highly directional and location-restricted procedure for generating clone contigs: a great deal of effort is invested in generating a contig from a fixed starting point on a specific chromosome, often as a prelude to positional cloning. In order to build clone contigs spanning large amounts of a chromosome, or even a whole genome, a more general, labour-efficient mapping approach is needed. To meet this need, a variety of clone fingerprinting techniques have been used to type clones at random and then integrate the information in order to identify clone contigs over large regions of a genome. For example, this approach has been used to great effect in assembling YAC contigs over significant amounts of the human genome. Clone fingerprinting techniques should be simple and rapid, and the methods described below have been widely used.

Repetitive DNA Fingerprinting

Mammalian DNA has a very high component of highly repeated, interspersed DNA sequences such as the human *Alu* and LINE-1 (*Kpn*) repeats (occurring on average about once every 3 kb and 10 kb respectively). Because the spacing between such interspersed repetitive DNA elements will vary in different genomic regions, clones which contain overlapping DNA sequences may be identified by similar patterns following hybridisation of repetitive DNA probes to Southern blots of the clone DNA digested with a suitable restriction endonuclease. For example, a whole genome clone fingerprinting approach has been applied to mapping the human genome largely on the basis of repetitive DNA fingerprinting of YACs (Bellanne-Chantelot *et al.*, 1992).

Bibliography

Ambrose, E. J., and Easty, D. M.: *Cell Biology: Reading*, Addison- Wesley, Mass, 1970.

Barrington, E. J. W.: *Biochemistry of Primitive Deuterostomians*, London, Academic Press, 1974.

Bauer, M.W., and G. Gaskell : *Biotechnology: The Making of a Global Controversy*, Cambridge: Cambridge University Press, 2002.

Bowler, Peter J.: *The Mendelian Revolution: The Emergence of Hereditarian Concepts in Modern Science and Society*. Baltimore: Johns Hopkins University Press, 1989.

Broach, J.R. : *The Molecular Biology of the Yeast,* Cold Spring Harbor: Cold Spring Harbor Laboratory, 1981.

Carroll SB, Grenier J, Weatherbee SD: *From DNA to Diversity: Molecular Genetics and the Evolution of Animal Design*. Oxford: Blackwell Publishing, 2005.

Chauhan, B.S. : *Principles of Biochemistry and Biophysics,* Laxmi Publications, Delhi, 2008.

Dhruve, J J : *Biochemistry at a Glance*, Agrotech Pub, Delhi, 2007.

Friedberg, E.C. : *DNA Repair,* New York: WH Freeman and Company, 1985.

Ganguly, Smriti : *Biochemistry of Biomolecules,* Pearl Books, Delhi, 2007.

Harris, John : *Wonderwoman and Superman: The Ethics of Human Biotechnology,* Oxford: Oxford University Press, 1992.

Hartl D, Jones E: *Genetics: Analysis of Genes and Genomes*. Jones & Bartlett, 2005.

Kitcher, Philip : *The Lives to Come: The Genetic Revolution and Human Possibilities*, New York: Simon and Schuster, 1996.

Lata Bhattacharya: *Biochemistry of Nutrition*, Discovery, Delhi, 2010.

Lodish, Harvey : *Molecular Cell Biology*, New York: W. H. Freeman, 2000.

Morholt, E. and Brandwein, P.F. : *Sourcebook for the Biological Sciences,* San Diego: Harcourt Brace Jovanovich, 1986.

Morris, Jonathan : *Ethics and Biotechnology*, New York: Chelsea House, 2005.

Narvekar, Raghunath : *Molecular Biochemistry : Principles and Practices,* Adhyayan Pub, Delhi, 2008.

Patel, S V ; U K Kandoliya: *Biochemistry at a Glance*, Agrotech Pub, Delhi, 2007.

Primrose, S. B. : *Molecular Biotechnology,* Boston: Blackwell Scientific Publications, 1991.

Reuter, Lars : *Modern Biotechnology in Postmodern Times? A Reflection on European Policies and Human Agency,* Dordrecht, Netherlands: Springer, 2003.

Sarath Chandra Bose: *Biochemistry : A Practical Manual*, Pharma Med Press, Delhi, 2010.

Sharma, Pradeep: *Biochemistry and Organisation of Cells*, RBSA Pub, Delhi, 2006.

Singh, Bimla: *Biochemistry and Genomic Revolution*, Vista International Pub, Delhi, 2006.

Trevor H. : *Transforming Matter: A History of Chemistry from Alchemy to the Buckyball,* Baltimore: Johns Hopkins University Press, 2001.

William Alec : *The Chemical Industry*, London: Longman Group Limited, 1971.

Yadav, Neelam: *Biochemistry and Biotechnology : A Laboratory Manual*, Pointer, Delhi, 2007.

Index

P

Q

R

S

T

U

❑❑❑